To my students,

who taught me that the best way to learn something
is to be forced to explain why it breaks.

Published by Binary Horizon Academic, United States.

Companion Resources & Ordering Information

For lecture slides, curriculum resources, or textbook adoptions, please visit www.aymanalzaid.com or contact the author directly through the website's support channels.

FIRST EDITION | First Printing: 2026

ISBN-13: 979-8-951087-02-7 (Paperback)

ISBN-13: 979-8-951087-03-4 (eBook)

Suggested Citation: Alzaid, A. (2026). *Computational Thinking and Software Design with Java.* Binary Horizon Academic.

Table of Contents

Acknowledgments

This textbook grew out of classroom teaching, and the students in those classrooms are its most important contributors. Every confusion they expressed, every question they asked at the wrong moment, and every bug they found in live code demonstrations made this book sharper, more honest, and more useful than it would otherwise have been. They taught me what actually needs to be said, and what can be left unsaid.

I am grateful to my colleagues for their encouragement and for the many conversations about pedagogy, curriculum design, and the rapidly shifting role of artificial intelligence in computing education. Those conversations shaped the framework at the heart of this book.

I would also like to thank the Computing Alliance of Hispanic-Serving Institutions (CAHSI) and the broader computing education community for the ongoing work of making computer science accessible, rigorous, and welcoming to every student who arrives at the door.

Peer reviewers will be acknowledged in future printings as the review process is completed. Instructors who adopt this text and share feedback are contributing to that process, and I am grateful in advance.

Ayman Alzaid
2026

Foreword

Computer science education is changing. The field no longer rewards students who can recite syntax from memory. It rewards students who can read unfamiliar code, spot the flaw in a logical argument, and reconstruct a broken method from first principles. These are not the skills a passive textbook produces.

This textbook does not ask you to memorize. It asks you to think. Every chapter is built on one conviction: the fastest path to genuine programming fluency is active, analytical engagement with code that challenges you, breaks your assumptions, and forces you to reason carefully about what the compiler and the Java Virtual Machine are actually doing.

Programming is the discipline of translating abstract human intent into precise, executable instructions. The gap between a beginner and a working engineer is not vocabulary. It is the habit of rigorous logical thinking, and the judgment to evaluate code, including AI-generated code, critically. This book is built to close that gap.

Preface

The Mindset: Software Engineering

Most beginners judge their code by one standard: did it compile and run without crashing? If yes, the program feels finished. Software engineering rejects that standard entirely. A program that runs is not automatically a program worth keeping.

Software engineering is the professional practice of building code that is functional, readable, and maintainable over the long life of a project. Think of code the way an architect thinks of a blueprint. The blueprint must be accurate enough for any other trained engineer to walk into the project, read the design without confusion, and build the exact same system. Clarity is not optional. It is the core deliverable.

The technique that makes this possible is problem decomposition: take a large, intimidating problem, break it into the smallest logical pieces you can identify, and solve each piece independently. Writing code is the act of recording those solutions in a language the machine can execute. Break before build. That is the engineering mindset this book builds.

A Note from the Author: Pedagogical Philosophy

This book is built on active learning. Reading a perfectly functional program teaches you almost nothing useful. You can follow every line, nod along, and still be completely stuck the moment you face a blank file. That false confidence is the specific failure this text is designed to prevent.

Every theory section is immediately followed by hands-on exercises. You will investigate code that is intentionally broken. You will predict the output of programs designed to test the edges of your assumptions. You will diagnose and rewrite poorly structured code against professional engineering standards. Each task requires you to reason actively rather than read passively.

In my own university classrooms, these become genuinely competitive exercises. Students compile the code to check whether they were right. They debate execution pathways with each other. They treat debugging tasks as problems to solve rather than chores to complete. The engagement level shifts completely once the material demands active participation. This book brings that same structure to independent study.

The habits this book builds, decomposing problems before writing code, tracing execution by hand, and evaluating code critically rather than accepting it at face value, are not Java-specific. They are the foundational habits of professional software development in any language. This book teaches them through Java because Java's strictness and object-oriented structure make those habits visible and necessary from the very first chapter.

AI Resilience and Authentic Assessment

The rapid rise of generative artificial intelligence has reshaped what it means to assess programming ability in higher education. When working code can be produced from a prompt in seconds, a student's ability to write boilerplate from memory is no longer the relevant measure. The relevant measure is what they can do with code they did not write: read it critically, identify what is wrong with it, explain why it fails structurally, and redesign it to professional standards.

This textbook is built as a deliberate response to that shift. Every active learning challenge in these pages requires something that language models cannot do on a student's behalf: trace a variable through a live execution environment, identify the architectural flaw in code that compiles and runs correctly, or defend a refactoring decision against a specific design principle. These tasks demand authentic analytical reasoning. They cannot be completed by pasting a question into a chat window.

For educators, this creates a practical framework for maintaining academic integrity without monitoring software or proctoring tools. The exercises are not resistant to AI by restriction or lockdown. They are resistant because they require the kind of reasoning that only comes from working through the problem yourself.

A Note to the Developer

In professional software engineering, there is no answer key to consult when verifying a production system. The compiler and the test suite are the authority. This book contains no pre-written solutions, not as an oversight, and not as a problem to work around. It is the point. The process of predicting output, compiling the code, and reconciling the difference between what you expected and what actually happened is where the learning lives. When your prediction is wrong, trace the code again, compile it again, and find exactly where your mental model diverged from the JVM. That gap is the lesson.

About the Author

Ayman Alzaid, Ph.D. is a Professor of Computer Science. He holds a Ph.D. with an interdisciplinary focus spanning Computer Science, Curriculum and Instruction, and Learning Design and Technology, alongside an M.S. in Computer Science and a B.S. in Computer Science.

His research focuses on artificial intelligence, human-computer interaction, and CS curriculum design.

He serves as a faculty partner of the Computing Alliance of Hispanic-Serving Institutions (CAHSI) and is a contributor to the Math Snacks project, an initiative leveraging educational games and media to enhance math and technology literacy. Through his writing, research, and teaching, he continuously works to create more accessible, engaging, and innovative educational experiences.

How to Use This Book

Each chapter follows the same structure. A Discovery Challenge opens with a hands-on, pre-syntax problem that surfaces the core concept before any Java is introduced. A Predict, Observe, Explain (POE) section presents a concrete scenario that builds intuition about how Java actually behaves. Naming What You Already Did formalizes the concept you just discovered in precise vocabulary. The Syntax Bridge connects that concept to real, runnable Java code. Practice Problems close each chapter with exercises that require active reasoning. Chapters 6, 8, 9, 10, and 12 also include an Architectural AI Audit, a structured exercise in which you evaluate AI-generated code, identify its flaw, and redesign it, the same critical evaluation skill professional software engineers use in code review.

You need two things before you begin: a working Java Development Kit (JDK 17 or later) and a code editor. IntelliJ IDEA Community Edition is recommended. VS Code with the Java Extension Pack and Eclipse are both solid alternatives. Every prediction you make in the POE sections and every audit you complete should be verified by actually compiling and running the code. The compiler is the authority.

Textbook Structure

This book moves through five progressive sections. Each section builds directly on the concepts established before it.

Section	Chapters	Core Topics
Foundations of Computation	1-4	Computational Thinking, Program State, Conditional Logic, Loops
Working with Java	5-6	Objects and Scanner, Arrays and 2D Arrays
Object-Oriented Design	7-10	Class Design, Encapsulation, Inheritance, Polymorphism
Algorithms and Robustness	11-13	Recursion, Algorithm Analysis, Exception Handling
System Design Capstone	14	System Design, ArrayList, Manager-Class Architecture

A Note on This Book's AI-Augmented, AI-Resilient Framework

Throughout this book, certain exercises (the Architectural AI Audits, appearing in Chapters 6, 8, 9, 10, and 12) are built around a deliberate three-role sequence we call the Professional Loop:

1. **The Draftsman (AI).** An AI assistant generates a first-pass solution: code that is syntactically correct, often idiomatic-looking, and sometimes entirely convincing on a casual read. In several cases, this code compiles and runs without error. That is intentional. The goal is not to hand students obviously broken code, but code that requires real architectural judgment to evaluate.
2. **The Architect (Student).** The student is explicitly instructed not to ask AI to find the flaw. They alone review the Draftsman's work, identify which design principle it violates, and reason out a stronger structure. This is the role the book is training: a developer who treats AI output as a draft to be evaluated, not an answer to be accepted. Students feel empowered in this role because they are correcting the AI, not the other way around.
3. **The Workshop (Compiler).** The student implements their corrected design and uses the compiler to confirm it. It is worth being precise about what this step proves and what it does not: the compiler verifies that code is syntactically correct and behaves as expected. It cannot verify that a design is architecturally sound. That judgment remains the Architect's alone. The Workshop step closes the loop on correctness; it does not replace the reasoning that came before it.

This sequence appears five times across the book (Chapters 6, 8, 9, 10, and 12) to build an AI-Augmented learning experience, while the same underlying skill, treating a confident AI output as a hypothesis to verify rather than a conclusion to trust, is reinforced in brief form throughout the remaining chapters' practice problems to ensure students remain AI-Resilient. The result is a student who is never discouraged from using AI as a tool, and never permitted to use it as a substitute for their own judgment.

The Dual Demand: Protecting Integrity, Ensuring Readiness

The AI-Augmented, AI-Resilient framework is a direct response to the two most urgent priorities in computer science pedagogy today:

AI-Resilient Assessment Design (The Protection)

AI-Resilient Pedagogy: Designing assessments that protect critical thinking, ensuring students cannot easily bypass learning via AI shortcuts.

As AI language models grow more capable of generating convincing code, computer science instructors face mounting pressure to prevent students from simply copy-pasting assignments into an AI tool to bypass learning. The research literature is explicit on this point: the term of art "AI-resilient assessments" now describes coding exercises deliberately structured to resist this failure mode. This book adopts that same strategy across every practice problem and coding example: decomposition, tracing, and verification are inherent to the problem design itself, not merely bolt-on instructions, producing exercises that are not just academically rigorous but cheat-resistant by construction.

AI-Augmented Learning (The Skill)

AI-Augmented Pedagogy: Deliberately weaving AI into tasks to enhance human capabilities and share the cognitive load.

Simultaneously, both academic and industry observers are clear: artificial intelligence is not a passing fad but a permanent fixture of the professional programmer's toolkit. The term "AI-augmented software engineering" describes a workflow this book is designed to instill as a habit: using AI as a partner in the act of creation, but never as the director of intent. Every exercise in this text steers the student toward viewing AI as a source of suggestions to be evaluated, incorporated, or discarded according to the student's own design judgment, never as a voice of authority. The result is a learner who gains the full productive benefit of AI-augmented coding without developing a dependence on it, the exact skill set the software industry now demands.

A resilient framework that protects foundational programming skills from AI reliance, while seamlessly preparing students for an AI-augmented software industry.

Aligned with AP Computer Science A (Fall 2025 CED) and ACM/IEEE CS1 guidelines. Designed for both AP and university-level introductory CS courses.

Supporting evidence: The term "AI-resilient assessments" appears in peer-reviewed computing education literature as a theoretically grounded and empirically validated framework, documenting a 30-percentage-point gap between AI-assisted homework performance and proctored exam results (Ding, 2024, arXiv:2512.10758; Akbar, 2025, arXiv:2503.23622). Industry adoption data: 85% of professional developers now regularly use AI coding tools, with 51% using them daily (JetBrains State of Developer Ecosystem, 2025; Stack Overflow Developer Survey, 2025; Modall, 2025).

A Note on Industry Readiness

This book is designed to produce not just students who can pass an exam, but developers who are ready to read and write professional-grade code from day one. Three specific design decisions reflect this goal.

Naming Conventions and Code Style

Every code example in this book follows Oracle's official Java code conventions and aligns with the Google Java Style Guide: class names in PascalCase (Student,

GraduateStudent, Roster), variables and methods in camelCase (getGpa(), addStudent()), and constants in SCREAMING_SNAKE_CASE (MAX_CAPACITY). Indentation uses 4 spaces throughout, the dominant convention in professional and teaching contexts. Students who graduate from this book will write code that is immediately readable by any professional Java developer.

One deliberate exception: loops in this book write i = i + 1 rather than the professional shorthand i++. This is a teaching scaffold only, making the increment operation fully explicit for a first-time reader. The instructor note in Chapter 4 explicitly acknowledges this and directs students toward i++ as the professional standard before they enter industry or open-source codebases.

Real-World Analogies Grounded in Production Practice

Every analogy in this book maps accurately to how the underlying concept behaves in production software, not just in introductory programs. The vending machine (Chapter 5) reflects how software engineers think about APIs: public interface, hidden implementation, no knowledge of internals required. The Roster/Student system (Chapters 7 through 14) is a realistic domain model of the kind found in real student information systems and enterprise applications. The Architectural AI Audit exercises (Chapters 6, 8, 9, 10, and 12) are modeled explicitly on the code review process used in professional software teams: one developer produces a draft (the Draftsman), a second evaluates it for correctness and design quality (the Architect), and the build system confirms correctness (the Workshop). Students who complete these exercises have practiced the most important professional software engineering skill that no compiler can teach: judgment about when code is correct but wrong.

Code Review Culture from Day One

The Architectural AI Audit exercises are the most direct industry preparation in this book. Code review is the single most universal professional software engineering practice, present in every team regardless of language, company size, or methodology. By introducing structured code evaluation in five separate chapters, each tied to a specific design principle the student has just learned, this book gives students their first experience of the evaluative mindset that distinguishes a professional engineer from someone who only writes code. @Override is required, not optional, in every code example that overrides a method, both because Oracle and Google style guides require it and because the habit of marking intent explicitly is itself a professional practice.

AP Computer Science A Alignment

This book is designed for both AP Computer Science A students and university-level CS1 courses. The table below maps each chapter to the AP CS A course framework (Fall 2025 CED, College Board). Chapters marked Beyond AP teach content outside the current AP exam scope, intentionally, because foundational

OOP concepts including inheritance and polymorphism are valuable preparation for CS2 and professional software development regardless of exam coverage.

Ch.	Chapter Title	Key Topics	AP Unit	Status
1	Computational Thinking	Decomposition, algorithm thinking, AI-Resilient framework	Unit 1	Covered
2	Program State	Variables, assignment, arithmetic, output	Unit 1	Covered
3	Conditional Logic	if/else, boolean expressions, relational operators	Unit 2	Covered
4	Process and Loops	while, for, loop tracing, iteration patterns	Unit 2	Covered
5	Working with Objects	Scanner, String methods (all AP QR methods), object creation	Unit 1	Covered
6	Data Collections	1D arrays, 2D arrays, traversal, ArrayIndexOutOfBoundsException	Unit 4	Covered
7	Designing Classes	Class design, constructors, CRC cards, switch/break/continue	Unit 3	Covered
8	Encapsulation	private fields, getters, setters, &&/||	Unit 3	Covered
9	Inheritance	extends, super, is-a relationship	Beyond AP*	Beyond AP
10	Polymorphism	@Override, polymorphic dispatch, instanceof	Beyond AP*	Beyond AP
11	Recursion	Base case, recursive case, StackOverflowError	Unit 4	Covered
12	Algorithm Analysis	Linear search, binary search, selection sort	Unit 4	Covered
13	Exception Handling	try/catch, do-while, Integer.parseInt, NumberFormatException	Unit 2+	Covered
14	System Design + ArrayList	Manager class design, ArrayList, for-each loop	Unit 3+4	Covered

** Chapters 9 and 10 cover inheritance and polymorphism, which were removed from the AP CS A exam in the Fall 2025 CED revision. These chapters are retained because foundational OOP knowledge remains essential for CS2 coursework and professional practice. Instructors using this book solely for AP exam preparation may treat these chapters as enrichment.*

References

Akbar, M. S. (2025). Beyond detection: Designing AI-resilient assessments with automated feedback tool to foster critical thinking and originality. University of Sydney. arXiv:2503.23622.

Ding, K. (2024). Designing AI-resilient assessments using interconnected problems: A theoretically grounded and empirically validated framework. University of Pennsylvania. arXiv:2512.10758.

JetBrains. (2025). State of developer ecosystem 2025. JetBrains Research.

Modall. (2025). AI in software development: 25+ trends and statistics. Retrieved from https://modall.ca/blog/ai-in-software-development-trends-statistics

Stack Overflow. (2025). Developer survey 2025. Stack Overflow.

A Note to Instructors

This book is designed for a one-semester introductory programming course using Java, at either the AP Computer Science A level or a university CS1 level. The fourteen chapters map to a standard sixteen-week semester with room for two review weeks, or can be compressed to twelve weeks by treating the companion sections (2D Arrays, ArrayList, do-while, and similar) as optional enrichment.

Pedagogical Design

Each chapter follows a fixed three-part structure: a Discovery Challenge (pre-syntax problem solving), a Logical Blueprint (conceptual mapping), and a Syntax Bridge (Java implementation). This structure is intentional and should not be skipped. The Discovery Challenge is where learning actually begins; students who skip it and read forward to the syntax consistently perform worse on conceptual questions than those who engage with it first.

The Architectural AI Audit exercises in Chapters 6, 8, 9, 10, and 12 are designed to be completed individually, not in pairs. The audit task requires the student to hold the full design in working memory and reason about it. These exercises are the most reliable AI-resilient assessments in the book: a student who cannot explain why the AI-generated code is wrong has not completed the task.

AI Policy and Academic Integrity

The AI-Resilient framework in this book is not a prohibition on AI tools. Students who use an AI assistant to generate the Discovery Challenge solution are bypassing the learning mechanism, not cheating in the traditional sense. The better framing for students is: the AI can produce the output, but only you can produce the understanding. The Confidence Checkpoint at the end of each chapter is designed to surface that distinction quickly.

Appendix C documents the AI-Augmented, AI-Resilient framework in detail for instructors who wish to cite it or adapt it for their own assessment design.

Companion Resources

Lecture slides are available at www.aymanalzaid.com. Instructors considering adoption are welcome to contact the author directly through the website for any questions about course integration. The Student Experience Survey (included separately) is a voluntary research instrument for instructors who wish to contribute outcome data to the ongoing evaluation of the AI-Augmented, AI-Resilient framework.

CHAPTER 1
Computational Thinking

An introduction to computational thinking, before any Java syntax.

THE SIGNPOST: Where We're Headed

In this chapter, we will learn how to break a big, confusing task into small pieces that are easy to handle, one at a time.

That's it. That's the whole goal of this chapter. Keep that one sentence in mind as you work through what's next. You don't need to know any Java yet to get started.

1.1 Discovery Challenge: The Robot Kitchen

Imagine you are giving instructions to a robot that is about to make a peanut butter and jelly sandwich. The robot is extremely literal. It does not assume anything. It does not know what "open" means unless you tell it exactly how. It cannot "use common sense."

Your task: write step-by-step instructions, in plain English, that the robot could follow exactly as written.

ONE RULE: Read this before you start

Your robot is not allowed to touch a lid, a jar, or a loaf of bread unless you have already given it a specific "Open" or "Retrieve" command first. No exceptions, no assumptions. If the robot hasn't been told to open the jar, the jar stays closed, even if the very next instruction is "spread the peanut butter."

TRY IT: Spend about 8 minutes on this

Grab a sheet of paper (or open a blank document). Write out every single step the robot needs, from an unopened bag of bread and jars on the counter, to a finished sandwich on a plate. Remember the one rule above.

It is completely normal not to finish, and it's even more normal to realize halfway through that you forgot a step (like opening the jar). That's expected. That's the point. We'll build the answer together right after this.

Most people, on their first try, write something like: "Put peanut butter on one slice, jelly on the other, put them together." That sounds complete, until the one rule kicks in. The jar is still closed. The bread bag is still sealed. Suddenly "put peanut butter on one slice" isn't one step at all. It's open the jar, retrieve the knife, scoop, spread, and only then move on.

This is the core discovery of this chapter: almost every task we think of as "one step" is secretly made of many smaller steps. A computer, like the robot, needs every one of those smaller steps spelled out. The skill of finding and ordering those smaller steps is the entire subject of this chapter.

1.2 Predict, Observe, Explain: The Repeating Step

Now suppose one of your robot's instructions is: "Spread peanut butter until the slice is covered." That single instruction is hiding something important: it's not one action, it's an action that repeats. Spread a little, check if it's covered, spread more if not, check again, until some condition becomes true.

PREDICT

Before reading further: what do you think would happen if the robot's instruction said "Spread peanut butter" with no condition at all for when to stop?

OBSERVE

Without a stopping condition, the literal-minded robot has no reason to ever stop spreading. It would keep going, forever. In real programs, this exact situation has a name: an infinite loop. It's one of the most common bugs beginning programmers run into, and you just discovered why it happens without writing a single line of code.

EXPLAIN

In your own words, jot down: why does a repeating instruction need a clear stopping condition? What goes wrong if it doesn't have one?

Notice what just happened. Your sandwich instructions couldn't really be "broken down" properly until you had a way to repeat a small action safely. That tool, repeating an action until a condition is met, has its own full name and its own chapter later in this book (Chapter 4). For now, you only need to recognize that the tool exists and that it needs a stopping condition. We are not mastering it yet. We are simply noticing that we need it.

Here is that same idea drawn out as a flowchart, a simple way to picture logic before writing any code. The diamond shape is a question with two possible paths.

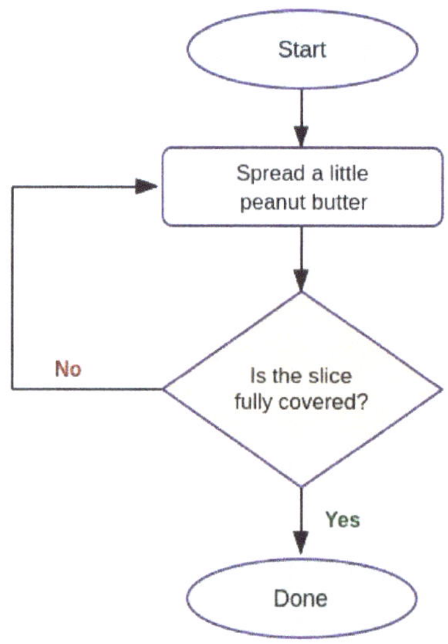

This loop only stops because of the "Is it covered?" check. Without it, the robot spreads forever.

You'll see this same diamond-and-arrow shape again throughout this book, any time we need to plan repeating logic before turning it into code.

1.3 Naming What You Already Did

FORMAL CONCEPT

The process you used on the sandwich problem, taking one big, vague task and splitting it into small, exact, ordered steps, is called decomposition.

You did not need to know that word to do it. You already had the instinct. Programmers have been using decomposition for decades, under that exact name, to take overwhelming problems and turn them into something a computer (which, remember, is just as literal as our robot) can actually carry out.

Every program you will ever write, from a simple calculator to a video game, starts with this same move: take the big task, break it into smaller tasks, and keep breaking those down until each piece is small enough to describe exactly.

1.3.1 Decomposition's Three Companions

Decomposition is one of four habits of mind that computer scientists call computational thinking, ways of approaching a problem that apply whether or not you're writing code at all. You've already brushed against the other three in this chapter too, even though none of them have been named yet:

- Decomposition: breaking a big problem into smaller, manageable pieces. This is what you just did with the sandwich.
- Pattern Recognition: noticing when a new problem is similar to one you've already solved. You'll do this constantly throughout this book: almost every chapter from here on will ask you to compare a new idea to something from an earlier chapter, because spotting that resemblance is most of the work.
- Abstraction: focusing on the details that matter for the task at hand and ignoring the ones that don't. The robot didn't need to know how bread is baked to make a sandwich. It only needed to know it had bread. You'll meet this idea again, by name, in Chapter 5.
- Algorithmic Thinking: designing a precise, repeatable, step-by-step procedure to solve a problem, not just a single answer. The numbered list you wrote for the sandwich was an early algorithm; Chapter 12 returns to this idea directly, comparing several algorithms for the same task.

You won't see these four terms drilled or tested individually. They're not vocabulary to memorize, they're a lens. Once you notice them, you'll start seeing all four at work in nearly everything this book asks you to do.

1.3.2 About This Language: Java

Before you see your first line of code, it's worth knowing a little about the language you're about to learn. Java was created in the early 1990s by James Gosling and his team at Sun Microsystems. It wasn't originally built for the kind of programs you'll write in this book, it started as a language for consumer electronics, like set-top boxes and televisions, but it was quickly repurposed for general software, and has remained one of the most widely used languages in the world ever since.

Two things about Java are worth knowing upfront, because they shape almost everything you'll see in this book:

- Java is platform-independent. A Java program can run, unchanged, on Windows, macOS, Linux, or almost any other system, because it compiles to an in-between format (called bytecode) rather than directly to one specific machine's instructions. This is often summarized as “write once, run anywhere.”
- Java is object-oriented. Almost everything in a Java program is built around objects, bundles of information and behavior grouped together. You'll start using objects someone else designed in Chapter 5, and start designing your own starting in Chapter 7, which is where the bulk of this book's design thinking lives.

Java is also known for being strict: it requires you to be precise about types, syntax, and structure in ways some other languages don't. That strictness can feel unforgiving at first, but it's also what catches a large category of mistakes before your program ever runs. You'll see this firsthand starting in Chapter 8, when private fields and compiler errors begin doing real protective work for you.

1.3.3 A Third Thread: AI-Augmented, AI-Resilient

The four computational thinking habits above are the first thread running through this book. Java's design and syntax is the second. There is a third, and it is worth naming now so you recognize it each time it appears.

Across five chapters in this book (6, 8, 9, 10, and 12), you will encounter a section called the Architectural AI Audit. Each one follows the same three-step Professional Loop:

- The Draftsman (AI): an AI assistant generates a working solution. The code compiles. It may even look reasonable. Your job is not to run it and accept it.
- The Architect (You): without asking AI to evaluate its own output, you read the code, find the flaw, and design something better. This is the skill the audit is training.
- The Workshop (Compiler): you implement your improved design and use the compiler to confirm it works. The compiler is the neutral judge, not the AI.

Beyond those five audit chapters, every practice problem section in this book ends with an optional reminder to apply the same habit to any AI-generated answer you used there too. The wording changes each time: tracing state in Chapter 2, tracing branches in Chapter 3, checking loop boundaries in Chapter 4, and so on. Same skill, same reason, different chapter vocabulary.

There is a name for what this trains. Using AI as a capable starting point for your work is called AI-Augmented practice. Treating AI output as a hypothesis to verify rather than a conclusion to trust is called AI-Resilience. Both are skills the software industry now expects from professional developers. This book builds them deliberately, in parallel, from this chapter forward.

ONE RULE: before you move on

This book never tells you not to use AI. It tells you to be the Architect, not the Draftsman. The difference is not which tools you use. It is who is directing the work.

1.4 Syntax Bridge: Your First Java Program

Let's connect this idea to real Java code for the first time. Below is one of the smallest possible Java programs. Its only job is to print a message to the screen.

```java
public class Greeting {
    public static void main(String[] args) {
        System.out.println("Hello, world!");
    }
}
```

Here is what each piece means, in plain English:

`public class Greeting` = *this names a container called "Greeting" that will hold our code*

`public static void main(` = *this marks the exact spot where the program starts running*

`String[] args)` = *a list of text the program could receive when it starts (we won't use this yet)*

`System.out.println(` = *print whatever is in the parentheses to the screen, then move to a new line*

`"Hello, world!"` = *the exact text we want printed, in quotes*

`;` = *marks the end of one instruction, every full instruction in Java ends this way*

Notice that even this tiny program is already decomposed: a container (the class), a starting point (main), and one instruction inside it (println). You'll see this same shape again: container, starting point, instructions, in almost every Java program in this book.

1.5 Confidence Checkpoint

Quick gut-check before moving on. These aren't graded, they're just for you, to see what's already sticking.

CHECK YOURSELF

1. In your own words, what is decomposition?
2. Why does a repeating instruction need a stopping condition?
3. In the Java program above, which line is where the program actually starts running?

If any of these feel shaky, that's useful information, not a failure. Flip back to that section before moving to Chapter 2.

Instructor Note: *Iteration appears here only as friction inside the POE. It is intentionally not taught in full. Its complete treatment, including for-loops and while-loops in Java syntax, is reserved for Chapter 4. Students are only expected to recognize the need for repetition and stopping conditions at this stage.*

CHAPTER 2

Program State and Memory

Giving the Computer Memory

THE SIGNPOST: Where We're Headed

In this chapter, we will learn how to give the computer something to remember, so it can use that information again later in the program.

In Chapter 1, every instruction ran once and was forgotten. Real programs need to hold onto values: a score, a name, a total, and use them again a few lines later. That's what this chapter is about.

2.1 Discovery Challenge: The Sticky Note Problem

Imagine you're running a lemonade stand with a friend, and you're keeping track of money entirely in your head, no paper, no notes. A customer pays you $5 for a $3 cup of lemonade. You need to remember: how much change do you owe them? How much total money is in the cash box now? How many cups have you sold today?

Now imagine ten customers in a row. Try to keep all three running numbers, cash box total, cups sold, change owed, in your head at once, updating them after every single customer, with no way to write anything down.

ONE RULE: Read this before you start

You are not allowed to write anything down, and you are not allowed to recalculate a number from scratch. If you forget a running total, it's gone, you'd have to guess. The only way to “remember” a number is to consciously hold it and update it, one step at a time.

TRY IT: Spend about 5 minutes on this

On paper, simulate five customers buying lemonade at $3 a cup, paying with different bills ($5, $10, $3 exact, $20, $5). After each customer, say out loud (or write only the running totals, not your working-out) what the cash box total and cups-sold count are. Notice how often you have to consciously hold onto a number just to use it one step later.

Here's the catch most people hit almost immediately: by the third or fourth customer, you're not really doing new math. You're doing old math plus new math, and the "old math" part is just last cycle's stored total. Without some place to keep that number, you'd have to recompute the entire day's sales from the very first customer, every single time.

This is the core discovery of this chapter: a running program is constantly doing the same thing you just did, holding onto values it computed earlier, so it doesn't have to redo the work. A computer needs a dedicated, named place to put each of those values. That named place is what this chapter is about.

2.1.1 Worked Example: Tracking the Lemonade Stand by Hand

Here's one way to track the five customers from the Discovery Challenge, written out explicitly so you can see exactly which numbers are being "remembered" and updated at each step.

- Start: cash box = $0, cups sold = 0
- Customer 1 pays $5 for $3 cup → cash box becomes $0 + $5 = $5, change owed = $2, cups sold becomes 1
- Customer 2 pays $10 for $3 cup → cash box becomes $5 + $10 = $15, change owed = $7, cups sold becomes 2
- Customer 3 pays exact $3 → cash box becomes $15 + $3 = $18, change owed = $0, cups sold becomes 3

Notice the pattern in every line: take the value you remembered from last time, combine it with something new, and store the result back in the same place, overwriting the old value. That "same place, new value" pattern is exactly what a variable does in Java.

2.2 Predict, Observe, Explain (POE): The Forgetful Cashier

Now imagine a cashier who, after every single customer, completely forgets the cash box total and has to ask the previous customer to remind them before they can continue.

PREDICT

Before reading further: what do you think happens to a lemonade stand run by a cashier who can't remember the running total between customers?

OBSERVE

The stand effectively can't function past the first customer. Every later step depends on a number from an earlier step. In a real Java program, this exact situation happens when a value is never stored anywhere. It's calculated once and then lost forever, unusable on the very next line.

EXPLAIN

In your own words, jot down: why does a program need a dedicated place to store a value, rather than just calculating it fresh every time it's needed?

This is also why programmers use the word state to describe these stored values. The current state of your lemonade stand at any moment is just the current values of cash box, cups sold, and change owed. A program's state is simply everything it currently remembers.

2.3 Naming What You Already Did

FORMAL CONCEPT

A named, reusable place to store a value, one that can be updated as the program runs, is called a variable. The act of holding values like this over time is what gives a program its state.

You already used variables, informally, every time you wrote “cash box becomes $5” in the worked example. You gave a quantity a name (“cash box”) and updated the value stored under that name. Java works exactly the same way, it just requires you to be precise about the name and what kind of value lives there.

2.4 Syntax Bridge: Declaring and Updating Variables

Here's the lemonade stand's cash box, written as a real Java program.

```
public class LemonadeStand {
    public static void main(String[] args) {
        int cashBox = 0;
        int cupsSold = 0;

        cashBox = cashBox + 5;
        cupsSold = cupsSold + 1;

        System.out.println(cashBox);
        System.out.println(cupsSold);
    }
}
```

🔍 **SYNTAX GLOSSARY**

Decoding the Java on this page, piece by piece

`int cashBox = 0;` = *create a named storage spot called cashBox, holding a whole number, starting at 0*

`int` = *this storage spot will only ever hold a whole number (no decimals)*

`cashBox = cashBox + 5;` = *take the current value of cashBox, add 5, and store the result back into cashBox*

`System.out.println(cashBox);` = *print whatever value is currently stored in cashBox*

Read that fifth line again slowly: cashBox = cashBox + 5. This looks like nonsense as a math equation. No number equals itself plus five. But it isn't math, it's an instruction: "take whatever is currently stored in cashBox, add 5 to it, and store that new result back in the same spot." This is the exact same move as "cash box becomes $5" from the worked example, just written in Java's precise grammar.

Notice also that cashBox and cupsSold are each declared once (with int in front, the very first time) and then updated freely afterward without int in front again. The first line creates the named spot; every line after that just uses it.

2.4.1 Worked Example: Tracking All Five Customers in Java

Here's the full Discovery Challenge, all five customers, written entirely in Java, using the same pattern.

```
public class LemonadeStandFull {
    public static void main(String[] args) {
        int cashBox = 0;
        int cupsSold = 0;

        cashBox = cashBox + 5;   // Customer 1 pays $5
        cupsSold = cupsSold + 1;

        cashBox = cashBox + 10;  // Customer 2 pays $10
        cupsSold = cupsSold + 1;

        cashBox = cashBox + 3;   // Customer 3 pays exact $3
        cupsSold = cupsSold + 1;

        System.out.println(cashBox);
        System.out.println(cupsSold);
    }
}
```

🔍 **SYNTAX GLOSSARY**

Decoding the Java on this page, piece by piece

`// Customer 1 pays $5` = *a comment, plain English notes for the reader, ignored entirely by Java*

By the end of this program, cashBox holds 18 and cupsSold holds 3, exactly matching your by-hand tracking from section 2.1.1. The computer didn't do anything you couldn't do; it just never forgets a value once it's stored, and never gets tired of repeating the same update pattern.

2.5 Practice Problems

Try these by hand first, the way you tracked the lemonade stand, then, if you'd like, try writing the Java version using the pattern from section 2.4.

1. Trace through a cash box that starts at $0 and receives three payments of $4, $6, and $2. Write the cash box value after each payment, the way section 2.1.1 did.
2. A program tracks pagesRead, starting at 0. You read 12 pages, then later that day 8 more pages. Write out what pagesRead equals after each reading session.
3. Explain, in your own words, why "x = x + 1" makes sense as a programming instruction even though it would never be true as a math equation.
4. Look back at the Forgetful Cashier POE. Name one real-world device or app you use that would completely fail to work if it had no memory of its previous state (for example, a step counter, a video game score, a shopping cart).

Optional: if you used an AI assistant on any of these, trace the variable's value by hand afterward, the way you traced cashBox above. An AI's confidence that a value is correct is not the same as it actually being correct.

2.6 Chapter Summary

KEY TAKEAWAYS

A variable is a named, reusable place to store a value that can be updated as the program runs.

A program's state is simply the current values of all its variables at any given moment.

Writing x = x + 1 means: take the current value of x, add 1, and store the result back into x. It is an instruction, not a math equation.

A variable is declared once (with its type, like int) and can be updated many times afterward without redeclaring it.

2.7 Confidence Checkpoint

Quick gut-check before moving on. These aren't graded, they're just for you, to see what's already sticking.

CHECK YOURSELF

1. In your own words, what is a variable?
2. Why does cashBox = cashBox + 5 make sense as an instruction, even though it isn't true as math?
3. What does it mean to say a program has "state"?

If any of these feel shaky, that's useful information, not a failure. Flip back to that section before moving to Chapter 3.

Instructor Note: *This chapter intentionally limits itself to the int type and simple arithmetic updates. Other primitive types (double, boolean, char), type casting, and naming conventions are addressed in the following chapter's Light Arc treatment. Conditional updates (only adding to cashBox if a condition is met) are deliberately deferred to Chapter 3: Conditional Logic, to avoid mixing two new concepts in one chapter.*

CHAPTER 3

Selection Structures

Teaching the Computer to Decide

THE SIGNPOST: Where We're Headed

In this chapter, we will learn how to make the computer choose between different actions, depending on what's true at the moment it runs.

Every program you've written so far in this book runs the exact same way, every single time. This chapter changes that. Your programs are about to start behaving differently depending on the situation.

3.1 Discovery Challenge: The Bouncer at the Door

Imagine you're working the door at a community event. Your job is simple: let people in if they meet the requirements, and turn them away if they don't. The requirement tonight is just one rule: a guest must be on the guest list.

Write out, in plain English, the exact instructions you'd give a brand-new, literal-minded trainee bouncer who has never worked a door before, someone who will follow your instructions exactly as written, with no judgment calls of their own.

ONE RULE: Read this before you start

Your instructions must cover both outcomes completely. "Let them in if they're on the list" is not finished. You have not told the trainee what to do in the other case. A literal-minded trainee left with no instruction for "not on the list" will simply freeze, unsure what to do next.

TRY IT: Spend about 5 minutes on this

Write the trainee's instructions for one guest arriving at the door. Cover both the "on the list" case and the "not on the list" case completely. Don't leave either one unfinished.

Most people write something close to: “Check the list. If their name is there, let them in. If their name is not there, turn them away.” Notice the shape of that sentence. It isn't one instruction, it's a fork: check one thing, and depending on the answer, go down one of exactly two different paths.

This is the core discovery of this chapter: almost every real decision a program needs to make has this same fork shape, a question with a yes path and a no path, both of which must be accounted for. Leaving either path unhandled is exactly like leaving the trainee bouncer with no idea what to do.

3.1.1 Worked Example: The Bouncer's Full Instructions

Here's one complete version of the trainee's instructions, written out so both paths are fully covered.

- Ask the guest for their name.
- Check the name against the guest list.
- If the name is on the list: let the guest in.
- If the name is not on the list: politely explain they are not on tonight's list, and direct them to the box office.

Notice that exactly one of those last two bullets will happen for any given guest, never both, never neither. That's the defining feature of this kind of fork: the two paths are mutually exclusive, and together they cover every possibility.

3.1.2 Worked Example: When There's More Than One Door Policy

Now suppose the event has a second rule: guests under 18 may enter, but only with a parent or guardian. The simple two-path fork from before isn't enough anymore. There are now three real outcomes to handle: on the list and an adult, on the list and a minor with a guardian, and not on the list at all.

- Check the name against the guest list.
- If the name is not on the list: turn the guest away.
- If the name is on the list AND the guest is 18 or older: let them in.
- If the name is on the list AND the guest is under 18: check for a parent or guardian. If one is present, let them both in. If not, turn the guest away and explain why.

Notice this is still the same fork instinct from section 3.1.1, just applied more than once, with one decision nested inside another. This “decision inside a decision” pattern is one you'll use constantly once you start writing real conditional logic.

3.2 Predict, Observe, Explain (POE): The Case Nobody Considered

Imagine a different trainee bouncer who was only given these instructions: “If the guest is on the list, let them in. If the guest is rude, turn them away.”

PREDICT

Before reading further: a polite guest arrives who is not on the list. What does this trainee do?

OBSERVE

Nothing in the instructions covers this guest at all. They're not rude, so the second rule doesn't apply, but they're also not on the list, so the first rule doesn't let them in either. The trainee has no instruction to follow. In a real program, this exact gap, a condition nobody accounted for, is one of the most common sources of bugs: code that behaves correctly for the cases the programmer thought of, and unpredictably for the ones they didn't.

EXPLAIN

In your own words, jot down: why is it not enough to only write instructions for the cases you expect to be common? What's the risk of leaving a case unhandled, even if you think it's rare?

This is exactly why your sandwich-and-bouncer instinct from section 3.1 mattered: covering both paths of a fork completely, not just the path you expect to happen most often.

Here is the bouncer's decision drawn as a flowchart, the same diamond-and-arrow style from Chapter 1, except this time, the diamond represents a one-time check rather than a repeating loop.

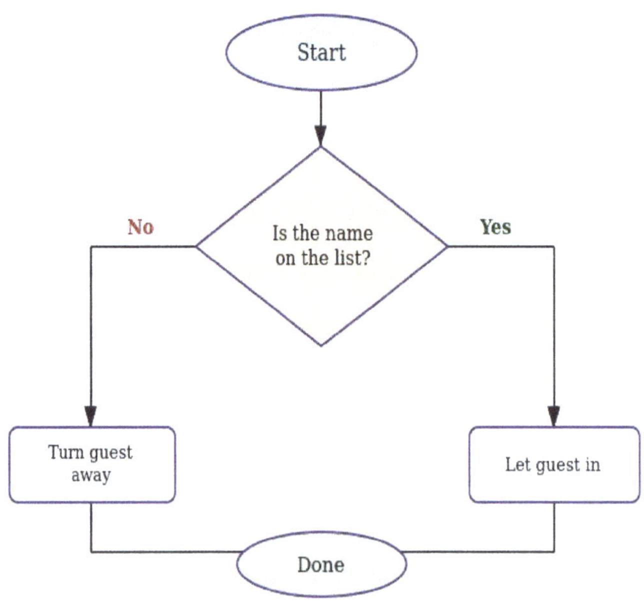

Compare this shape to the repeating loop flowchart from Chapter 1. There, the No path looped back up to repeat a step. Here, both paths lead forward to Done. A decision happens once, then the program moves on. Same diamond shape, two very different jobs.

3.3 Naming What You Already Did

FORMAL CONCEPT

Choosing between two or more different actions, based on whether something is true or false at the moment the program runs, is called conditional logic, and the specific yes/no fork you just designed is called a branch.

You didn't need either term to design the bouncer's instructions. You already had the instinct: ask a yes/no question, then do one of two different things depending on the answer, and make sure both answers are accounted for. Java gives this exact instinct a precise grammar.

3.4 Syntax Bridge: if and else

Here's the simple version of the bouncer, written as a real Java program.

```
public class DoorCheck {
    public static void main(String[] args) {
        boolean onTheList = true;

        if (onTheList) {
            System.out.println("Welcome in!");
        } else {
            System.out.println("Sorry, you're not on tonight's list.");
        }
    }
}
```

🔍 **SYNTAX GLOSSARY**

Decoding the Java on this page, piece by piece

`boolean onTheList = true;` = *create a storage spot called onTheList that holds only true or false*

`boolean` = *a type that can only ever be true or false, nothing else*

`if (onTheList) {` = *if the value in onTheList is true, run the instructions inside these braces*

`} else {` = *otherwise (if onTheList is false), run these instructions instead*

Notice the shape: if (...) opens one path, else opens the other, and exactly one of the two blocks runs, never both, never neither. That's the exact fork you designed by hand in section 3.1, now written in Java's precise grammar.

Try changing onTheList to false in your head before reading on: which message prints? If you can answer that without running the code, you've already internalized how if/else works.

3.4.1 Worked Example: A Nested Decision in Java

Here's the more complete door policy from section 3.1.2, list status, plus an age and guardian check, written in Java. Notice the if/else inside another if/else, matching the "decision inside a decision" shape from the worked example.

```
public class DoorCheckFull {
    public static void main(String[] args) {
        boolean onTheList = true;
        int age = 16;
        boolean hasGuardian = true;

        if (onTheList) {
            if (age >= 18) {
                System.out.println("Welcome in!");
            } else if (hasGuardian) {
                System.out.println("Welcome in, with your guardian.");
            } else {
                System.out.println("Minors need a guardian to enter.");
            }
        } else {
            System.out.println("Sorry, you're not on tonight's list.");
        }
    }
}
```

🔍 **SYNTAX GLOSSARY**

Decoding the Java on this page, piece by piece

`age >= 18` = *a comparison: is age greater than or equal to 18? Result is true or false*

`} else if (hasGuardian) {` = *if the first condition was false, check this new condition before falling back to else*

Trace through this by hand with age = 16 and hasGuardian = true, the same way you traced cashBox in Chapter 2: onTheList is true, so we enter the outer if. age >= 18 is false (16 is not ≥ 18), so we check the else if: hasGuardian is true, so that branch runs, and "Welcome in, with your guardian." prints. Every other branch is skipped entirely.

3.5 Practice Problems

Try these by hand first, tracing through the logic the way you traced the door check above, then, if you'd like, try writing the Java version using the pattern from section 3.4.

5. Write plain-English if/else instructions for a vending machine: if the inserted amount is enough to cover the price, dispense the item; otherwise, return the money. Make sure both paths are fully covered.
6. Trace through the DoorCheckFull program from section 3.4.1 by hand using age = 20 and onTheList = true. Which line prints? Show your reasoning step by step.
7. Trace through DoorCheckFull again, this time with onTheList = false, age = 16, and hasGuardian = false. Which line prints? Does the value of age or hasGuardian matter at all in this case? Why or why not?
8. Look back at the POE in section 3.2. Describe, in your own words, a real-world example (not the bouncer) where an unhandled case in a decision could cause a real problem.

Optional: if you used an AI assistant on any of these, trace every branch by hand afterward, the same way you traced DoorCheckFull above. A confident-sounding answer is not the same as a fully accounted-for one.

3.6 Chapter Summary

KEY TAKEAWAYS

Conditional logic means choosing between different actions depending on whether something is true or false when the program runs.

A branch (if/else) must account for every possible case. An unhandled case is a common source of bugs.

if (condition) { ... } else { ... } runs exactly one of its two blocks, never both, never neither.

Conditions can be nested inside one another to handle more complex decisions, the same way the door policy combined list status, age, and guardian status.

3.7 Confidence Checkpoint

Quick gut-check before moving on. These aren't graded, they're just for you, to see what's already sticking.

CHECK YOURSELF

1. In your own words, what is a branch?

2. Why is it a problem to leave a possible case unhandled in an if/else, even if you think it's rare?

3. In an if/else statement, is it possible for both blocks to run? Why or why not?

If any of these feel shaky, that's useful information, not a failure. Flip back to that section before moving to Chapter 4.

Instructor Note: *This chapter treats if/else and the comparison operators (>=, ==, etc.) as an Anchor-level first encounter, deliberately limited to boolean and simple numeric comparisons. The switch statement is intentionally not covered here. It is a Light Arc addition once branching itself is no longer new, and is addressed in section 7.0, a short companion section at the start of Chapter 7. Logical operators (&&, ||, !) for combining multiple conditions in a single if are also deferred, to avoid layering too many new symbols onto students' first branching experience at once.*

CHAPTER 4

Looping Structures

Teaching the Computer to Repeat

THE SIGNPOST: Where We're Headed

In this chapter, we will learn how to make the computer repeat an action automatically, instead of writing the same instruction over and over by hand.

Back in Chapter 1, the sandwich instructions hit a wall: "spread until covered" needed repetition, and we set that need aside on purpose. This chapter is where we finally pick it back up, and this time, we build it properly, with a plan before any code.

4.1 Discovery Challenge: The Copy-Paste Problem

Suppose you want a Java program that prints "Happy Birthday!" five times, once per line. Using only what you learned in Chapters 1 through 3, no new tools, the only way to do this is to write System.out.println("Happy Birthday!"); five separate times, one after another.

```
public class Birthday {
    public static void main(String[] args) {
        System.out.println("Happy Birthday!");
        System.out.println("Happy Birthday!");
        System.out.println("Happy Birthday!");
        System.out.println("Happy Birthday!");
        System.out.println("Happy Birthday!");
    }
}
```

ONE RULE: Read this before you start

Now imagine the requirement changes: print it 100 times instead of 5. You are not allowed to use copy-paste, and you are not allowed to write more than 6 lines of code total. Think for a moment about whether the program above could ever satisfy that requirement, no matter how you rearranged it.

TRY IT: Spend about 5 minutes on this

On paper, sketch in plain English (not Java yet) how you would describe "print this line 100 times" as a small set of instructions, without writing the print line 100 separate times. What's the smallest number of distinct ideas you need to fully describe the task?

Most people land on something like: "Keep a count starting at 0. Print the line. Add 1 to the count. If the count is less than 100, do it again. Otherwise, stop." Notice that's only four distinct ideas: a starting point, an action, an update, and a check, no matter whether the task is 5 repeats or 5,000.

This is the core discovery of this chapter: any task that repeats doesn't need to be written out repeatedly. It needs four ingredients: a starting point, an action to repeat, an update, and a stopping check, combined into a single reusable structure. That structure is what this chapter builds.

4.2 Logical Blueprint: Planning the Repeat Before Writing Code

Before touching Java syntax, let's formalize the four ingredients from the Discovery Challenge as a flowchart, the same diamond-and-arrow style from Chapter 1's "spread until covered" loop, and structurally identical to it.

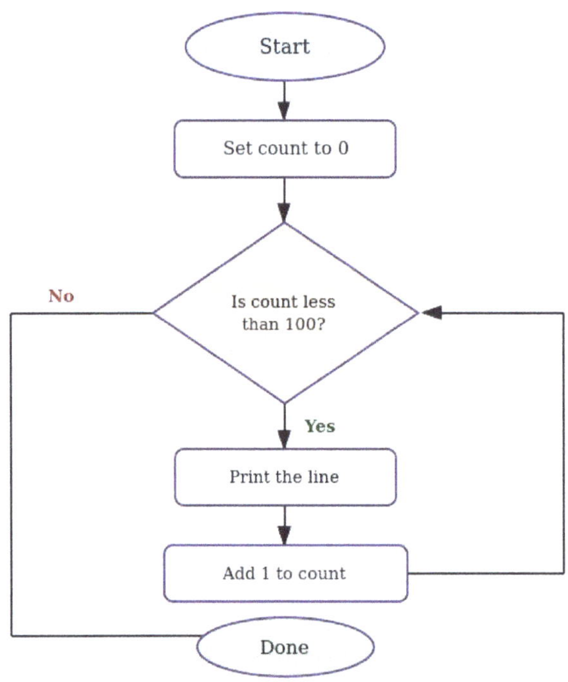

Compare this to the "spread until covered" flowchart from Chapter 1. Same shape, same four ingredients, just relabeled: a starting value (count = 0), an action (print the line), an update (add 1 to count), and a stopping check (is count less than 100?). Every loop you will ever write, in any language, has this same skeleton.

Notice something important about the order: the stopping check happens before the action runs, not after. That ordering choice matters. It's the difference between a loop that might run zero times and one that always runs at least once. We'll come back to this distinction later in the chapter.

4.2.1 Worked Example: From Flowchart to Pseudocode

Pseudocode is a way to write out a plan in structured, code-like English, without worrying about Java's exact grammar yet. Here's the flowchart above, written as pseudocode:

```
SET count TO 0
WHILE count IS LESS THAN 100
    PRINT "Happy Birthday!"
    SET count TO count + 1
END WHILE
```

Each line of this pseudocode maps directly onto one piece of the flowchart: SET count TO 0 is the starting point, WHILE … IS LESS THAN 100 is the stopping check, PRINT is the action, and SET count TO count + 1 is the update, the exact same update pattern (x = x + 1) you practiced in Chapter 2.

4.3 Syntax Bridge: The while Loop

Here's that pseudocode translated directly into Java. Read it side by side with the pseudocode above. The structure barely changes.

```java
public class BirthdayLoop {
    public static void main(String[] args) {
        int count = 0;

        while (count < 100) {
            System.out.println("Happy Birthday!");
            count = count + 1;
        }
    }
}
```

SYNTAX GLOSSARY

Decoding the Java on this page, piece by piece

`int count = 0;` = *the starting point, a variable to track how many times we've repeated so far*

`while (count < 100) {` = *the stopping check, keep repeating the block below as long as this stays true*

`count = count + 1;` = *the update, the same x = x + 1 pattern from Chapter 2, run once per repeat*

This six-line program does the work that would have taken 100 separate println lines using only Chapters 1–3 tools. The computer checks count < 100, runs the block if true, updates count, and checks again, automatically, without you writing the repeat by hand.

4.4 Predict, Observe, Explain (POE): Two Common Loop Bugs

Loops introduce two bugs that are extremely common for beginners and worth meeting on purpose, in a safe setting, before you meet them by accident in your own code.

PREDICT: Bug 1

Suppose the update line, count = count + 1;, was accidentally left out of the loop entirely, every other line stays the same. What do you predict happens when this program runs?

OBSERVE: Bug 1

count stays at 0 forever, since nothing ever changes it. The condition count < 100 is true on every single check, forever. The program prints “Happy Birthday!” endlessly and never reaches Done. This is the infinite loop you first predicted all the way back in Chapter 1's POE, except now you can see exactly which missing line causes it.

PREDICT: Bug 2

Now suppose the update line is back in place, but the condition is changed from count < 100 to count <= 100 (less than or equal to). How many times does the line print now? Still 100, or something else?

OBSERVE: Bug 2

It prints 101 times. count takes the values 0, 1, 2, ..., 100, and the check allows count = 100 to still pass, since 100 ≤ 100 is true. This single-character bug is common enough to have its own name: an off-by-one error. It's one of the most frequent mistakes in real, professional code, not just student code.

EXPLAIN

In your own words, jot down: why does forgetting the update line cause a completely different kind of bug than changing < to <=? What is each bug actually telling you about the loop's stopping condition?

4.4.1 Worked Example: The for Loop

Because the start-check-update pattern is so common, Java offers a second way to write the exact same loop, with all three of those pieces gathered into one line. This is called a for loop, and it does precisely the same thing as the while loop in section 4.3, just organized differently.

```
public class BirthdayForLoop {
    public static void main(String[] args) {
        for (int count = 0; count < 100; count = count + 1) {
            System.out.println("Happy Birthday!");
        }
    }
}
```

🔍 **SYNTAX GLOSSARY**

Decoding the Java on this page, piece by piece

`for ( ; ; ) {` = *a loop with all three parts, start, check, update, gathered into one line*

`int count = 0;` = *first part: runs once, before the loop starts (the starting point)*

`count < 100;` = *second part: checked before every repeat (the stopping check)*

`count = count + 1` = *third part: runs automatically after every repeat (the update)*

Nothing new is happening here conceptually. This is the identical four-ingredient loop from the flowchart in section 4.2, just written more compactly. Programmers reach for for when the number of repeats is known or countable in advance, and while when the stopping condition depends on something less predictable, like user input or a search result. You'll see both used throughout the rest of this book.

4.5 Refactor: Turning Repeated Code Into a Loop

Refactoring means rewriting working code to be cleaner or more general, without changing what it actually does. Here's a small program that prints the first five square numbers (1, 4, 9, 16, 25), written the "copy-paste" way from section 4.1, before you knew about loops:

```
public class SquaresCopyPaste {
    public static void main(String[] args) {
        System.out.println(1 * 1);
        System.out.println(2 * 2);
        System.out.println(3 * 3);
        System.out.println(4 * 4);
        System.out.println(5 * 5);
    }
}
```

Now here's the same program, refactored into a for loop using everything from this chapter:

```
public class SquaresRefactored {
    public static void main(String[] args) {
        for (int n = 1; n <= 5; n = n + 1) {
            System.out.println(n * n);
        }
    }
}
```

Both programs print the exact same five numbers. But the refactored version scales. Changing n <= 5 to n <= 5000 prints the first 5,000 squares with a one-character edit, while the copy-paste version would need 4,995 new lines written by hand. This is the real payoff of everything in this chapter: not just avoiding repetitive typing, but writing programs that stay easy to change.

4.6 Practice Problems

Try tracing these by hand first, the way you traced the door check in Chapter 3, then write the Java version using the patterns from this chapter.

- Write a for loop that prints the numbers 1 through 10, each on its own line. Trace through it by hand first: what is the value of your loop variable on the very last repeat?
- A while loop is meant to print "Loading..." ten times, but the programmer forgot the update line, exactly like Bug 1 in section 4.4. Explain what will happen if this code is run, and how you would fix it.
- Trace through a for loop with the condition count <= 5 instead of count < 5, starting at count = 0. List every value count takes, and state how many times the loop body runs.

- Refactor this copy-paste code into a single for loop: System.out.println("Lap 1 complete"); System.out.println("Lap 2 complete"); System.out.println("Lap 3 complete");

Optional: if you used an AI assistant on any of these, check its loop's boundary by hand afterward, the exact off-by-one trace from section 4.4. AI-generated loops get boundaries wrong often enough that this is worth doing every time.

4.7 Chapter Summary

KEY TAKEAWAYS

Every loop has four ingredients: a starting point, an action to repeat, an update, and a stopping check.

A while loop checks its condition before every repeat, including the very first one.

A for loop gathers the start, check, and update into a single line. It behaves identically to an equivalent while loop.

Forgetting the update line causes an infinite loop; a slightly wrong condition (like <= instead of <) causes an off-by-one error.

Refactoring repeated code into a loop doesn't change what a program does. It changes how easy the program is to extend later.

4.8 Confidence Checkpoint

Quick gut-check before moving on. These aren't graded, they're just for you, to see what's already sticking.

CHECK YOURSELF

1. In your own words, what are the four ingredients every loop needs?
2. What's the difference between an infinite loop and an off-by-one error? What causes each one?
3. When might you choose a for loop over a while loop, or vice versa?

If any of these feel shaky, that's useful information, not a failure. Flip back to that section before moving to Chapter 5.

Instructor Note: *This chapter deliberately covers while and for but omits do-while, which guarantees at least one execution regardless of the condition. do-while is intentionally introduced in section 13.3.2, alongside user-input validation patterns where its "run once, then check" behavior has a clear motivating use case, rather than here where it would be one more variant with no concrete reason to prefer it. Loop control statements (break, continue) and the switch statement are similarly deferred to section 7.0, a short companion section at the start of Chapter 7, where Chapter 6's arrays give break and continue a natural, motivated context. A note on loop increment style: this chapter writes i = i + 1 throughout to keep the operation fully explicit for first-time readers. Professional Java almost universally uses i++ or i += 1, which are exact shorthand equivalents. Students moving into industry codebases or open-source projects should adopt i++ as their default; the verbose form here is a teaching scaffold, not a style recommendation.*

CHAPTER 5

User Interaction and I/O

Using Tools Someone Else Built

THE SIGNPOST: Where We're Headed

In this chapter, we will learn how to use tools that someone else already built, without needing to know exactly how they work on the inside.

Every program you've written so far has been entirely your own code, from scratch. Real programs almost never work that way. They constantly borrow finished, tested tools other programmers already built. This chapter is about learning to use one of those tools well, before you ever try to build your own.

5.1 Discovery Challenge: The Vending Machine

Think about the last time you used a vending machine. You didn't need to know how the coin sensor works, how the internal motor turns to release a snack, or how the machine keeps track of inventory. You needed exactly three things: a way to put money in, a way to pick an item, and a way to get your item and change out.

Now imagine someone hands you a brand-new vending machine model you've never used before, with no instructions. List, in plain English, what you would need to figure out before you could successfully buy a snack. Then notice what you would NOT need to figure out.

ONE RULE: Read this before you start

You are not allowed to open the machine or look inside it. Everything you list must be something visible and usable from the outside only: buttons, slots, a screen, a door. If it requires opening the machine to know, it doesn't belong on your list.

TRY IT: Spend about 5 minutes on this

Write two short lists side by side: "What I need to know to use this machine" and "What I don't need to know to use this machine." Be specific. For the second list, think about everything happening mechanically inside that you've genuinely never needed to understand.

Your "need to know" list probably includes things like: which slot accepts money, which buttons correspond to which items, where the item comes out. Your "don't need to know" list is almost certainly much longer: the wiring, the motor mechanism, how change is calculated and dispensed, how the machine tracks what's sold out.

This is the core discovery of this chapter: a well-built tool draws a clear line between what you must know to use it, and everything else, which stays hidden and irrelevant to you as the user. That line isn't an accident. It's a deliberate design choice, and it's exactly how the tools you're about to use in Java were built.

5.1.1 Worked Example: Two Lists for a Vending Machine

Here's one reasonable version of the two lists from the Discovery Challenge.

What I need to know to use this machine:

- Which slot accepts money, and what forms of payment it takes
- Which button or code corresponds to the snack I want
- Where the snack and any change come out

What I don't need to know to use this machine:

- How the coin or card reader verifies payment
- How the internal spiral mechanism releases the chosen item
- How the machine calculates and dispenses exact change
- How the machine tracks which items are running low

Notice the shape of this split: the first list is short and stable. It rarely changes, even when the machine is upgraded internally. The second list could change completely (a new motor, a new payment chip) without you, the user, ever needing to notice or care.

5.2 Predict, Observe, Explain (POE): The Machine Gets an Upgrade

Suppose the vending machine company replaces the entire internal mechanism: new motor, new payment chip, new change-dispensing system, but keeps the buttons, the coin slot, and the item-pickup door exactly the same as before.

PREDICT

Before reading further: do you, as a returning customer, need to relearn how to use this machine? Why or why not?

OBSERVE

No. You use it exactly the same way you always have. The entire internal mechanism changed, but nothing about your experience as a user changed at all, because everything you relied on (the buttons, the slot, the door) stayed the same. This is true of well-built software tools too: the people who maintain Java's built-in tools can completely rewrite the internals for speed or correctness, and your code using those tools keeps working without any changes, as long as the outside, what you actually call and use, stays the same.

EXPLAIN

In your own words, jot down: why is it useful, as someone using a tool, for the inside of that tool to be allowed to change without affecting how you use it?

5.3 Naming What You Already Did

FORMAL CONCEPT

A pre-built tool that you can use through a known, stable set of buttons-and-slots, without needing to know how it works inside, is called an object. The specific actions an object lets you perform are called its methods.

You already sorted exactly this distinction, the "buttons and slots" versus "the hidden mechanism," for the vending machine. In Java, every object works the same way: it offers you a specific set of methods you can call, and everything happening behind those methods is none of your concern as the person using it.

This chapter is deliberately about using objects someone else carefully designed, not building your own yet. That comes in Chapter 7, once you've gotten comfortable on the "user" side of this line first.

5.4 Syntax Bridge: Using the Scanner Object

Every program you've written so far has used fixed, hardcoded values. cashBox always started at 0, the bouncer's list status was always set directly in the code. Real programs need to react to whatever the user actually types. Java provides a built-in object called Scanner specifically for this. It's the "vending machine" for reading keyboard input.

```
import java.util.Scanner;

public class GreetUser {
    public static void main(String[] args) {
        Scanner keyboard = new Scanner(System.in);

        System.out.println("What is your name?");
        String name = keyboard.nextLine();

        System.out.println("Hello, " + name + "!");
    }
}
```

SYNTAX GLOSSARY

Decoding the Java on this page, piece by piece

`import java.util.Scanner;` = *tells Java to make the Scanner tool available in this file, before we use it*

`Scanner keyboard = new Scanner(System.in);` = *create a Scanner object, hooked up to the keyboard, named keyboard*

`new Scanner(System.in)` = *the "button" that builds a brand-new Scanner object ready to use*

`keyboard.nextLine()` = *a method call, ask the keyboard object to wait for and return one line the user types*

`String name = …` = *store whatever the user typed into a variable called name, the same pattern from Chapter 2*

Notice keyboard.nextLine(). This dot-notation is how you call a method on an object: object name, a dot, then the method name and parentheses. You're pressing one of the Scanner object's "buttons." You don't need to know how nextLine() listens for keyboard input internally, exactly like the vending machine, that's intentionally none of your concern.

5.4.1 Worked Example: Methods on a String Object

You've been using String values since Chapter 1 ("Hello, world!") without knowing it, but String is itself an object, and it comes with its own set of useful methods, the same way Scanner does.

```java
public class StringMethods {
    public static void main(String[] args) {
        String name = "ada lovelace";

        System.out.println(name.length());
        System.out.println(name.toUpperCase());
        System.out.println(name.indexOf("love"));
    }
}
```

🔍 SYNTAX GLOSSARY

Decoding the Java on this page, piece by piece

`name.length()` = *a method call, ask the name object how many characters it holds*

`name.toUpperCase()` = *a method call, ask for an all-uppercase version of the text in name*

`name.indexOf("love")` = *a method call, ask for the position where "love" first appears inside name*

Each line follows the exact same dot-notation pattern as keyboard.nextLine(): object name, dot, method name, parentheses. You don't need to know how toUpperCase() converts letters internally, you only need to know it exists, what it's called, and what it gives back. That's the entire "need to know" list, just like the vending machine.

5.4.2 Companion Section: More String Methods

Three more String methods appear regularly in real programs and on the AP Java Quick Reference. Each one follows the same dot-notation pattern you already know.

```java
String word = "hello";

System.out.println(word.substring(1, 3));
System.out.println(word.charAt(0));

String other = "hello";
System.out.println(word.equals(other));
```

> 🔍 **SYNTAX GLOSSARY**
> *Decoding the Java on this page, piece by piece*
>
> `word.substring(1, 3)` = *returns the characters from position 1 up to but not including position 3, prints el*
>
> `word.charAt(0)` = *returns the single character at position 0, prints h*
>
> `word.equals(other)` = *returns true if both Strings hold the same text, false otherwise*

Notice that substring uses two position numbers: where to start (included) and where to stop (not included). This is the same 0-indexed, positions-start-at-zero rule from arrays in Chapter 6.

One important rule: never use == to compare two String values. Use equals() instead. The == operator checks whether two variables point to the exact same object in memory, which is not what you usually want. equals() checks whether the text inside both Strings is the same, which is what you actually mean when you ask whether two strings match.

> **ONE RULE: before you move on**
>
> word == other and word.equals(other) are not the same question. Use equals() whenever you want to compare String contents. This is one of the most common Java mistakes, and it produces no error message, just wrong behavior.

5.5 Practice Problems

Try these by hand first, predicting the output the way you did in earlier POE sections, then, if you'd like, run the Java versions to check yourself.

9. Write a short program that asks the user for their favorite color using Scanner, then prints it back in a full sentence.
10. Given String city = "Albuquerque";, predict the output of city.length() and city.toUpperCase() without running any code. Check your reasoning against what you learned about String methods in this chapter.
11. Look back at the vending machine's two lists from section 5.1.1. Make the same two lists for a Java object you've now used, Scanner, listing what you needed to know to use it, versus what stayed hidden.
12. In your own words, explain why import java.util.Scanner; is needed at the top of a file before keyboard.nextLine() can be used. Connect this to the vending machine's “buttons and slots” idea if it helps.

Optional: if you used an AI assistant on any of these, predict each method's output yourself before reading its explanation, the same predict-first habit from this chapter's POE sections. An AI explaining a method confidently doesn't mean it picked the right one.

5.6 Chapter Summary

KEY TAKEAWAYS

An object is a pre-built tool you can use through a known set of methods, without needing to know how it works inside.

A method call uses dot notation: objectName.methodName(), the same shape every time.

Scanner is a built-in Java object used to read input the user types; String is itself an object too, with its own useful methods like length() and toUpperCase().

This chapter is about using objects someone else designed. Chapter 7 is where you design your own.

5.7 Confidence Checkpoint

Quick gut-check before moving on. These aren't graded, they're just for you, to see what's already sticking.

CHECK YOURSELF

1. In your own words, what is an object?
2. What does objectName.methodName() actually do, in plain English?
3. Why didn't you need to know how Scanner reads keyboard input internally in order to use it?

If any of these feel shaky, that's useful information, not a failure. Flip back to that section before moving to Chapter 6.

Instructor Note: *This chapter deliberately avoids the words "class," "instance," and "constructor" as formal vocabulary, even though new Scanner(System.in) is technically a constructor call. The goal here is purely to build comfort calling methods on existing objects (the user role) before introducing class design vocabulary (the designer role) in Chapter 7. Revisit new Scanner(System.in) explicitly in Chapter 7 once "constructor" has been formally introduced, to connect the dots retroactively.*

CHAPTER 6

Data Collections

Managing Large Sets of Information

THE SIGNPOST: Where We're Headed

In this chapter, we will learn how to store many related values under a single name, instead of inventing a new variable for every single one.

Every variable you've used so far has held exactly one value: one cash box total, one name, one age. Real programs constantly need to hold many values of the same kind at once: every student's grade, every player's score, every item in a cart. This chapter is about doing that without your code falling apart as the amount of data grows.

6.1 Discovery Challenge: The Five Variables Problem

Suppose you're writing a program to average a small class's quiz scores. With five students, using what you already know from Chapters 2–4, you might declare: int score1, score2, score3, score4, score5;, fill each one in, add them all together, and divide by 5.

```
int score1 = 88;
int score2 = 92;
int score3 = 79;
int score4 = 95;
int score5 = 84;

int total = score1 + score2 + score3 + score4 + score5;
double average = total / 5.0;
```

ONE RULE: Read this before you start

Now imagine the class grows to 35 students, and by next semester, 200. You are not allowed to write 200 separate variable names, and you are not allowed to write a 200-term addition by hand. Think for a moment about whether the five-variable approach above could ever realistically scale to 200 students.

TRY IT: Spend about 5 minutes on this

On paper, sketch what would actually break first if you tried to scale the five-variable approach to 200 students. Then sketch, in plain English, what you'd want instead: a single named "thing" that could hold all 200 scores, that a loop (from Chapter 4) could walk through one at a time.

The five-variable approach breaks for a very specific reason: score1 through score200 are 200 completely unrelated names, as far as Java is concerned. There's no way to write a loop that visits "the next one," because nothing connects score47 to score48 except the human-readable pattern in their names, which Java cannot see or use.

This is the core discovery of this chapter: what you actually need is a single named container that holds many values, where each value can be reached by its position, first, second, third, and so on, so that a loop can visit every position automatically, the same loop working whether there are 5 values or 5,000.

6.2 Logical Blueprint: Positions, Not Names

Before touching Java syntax, let's formalize the idea from the Discovery Challenge. Instead of five separate names, picture one labeled row of boxes, each holding one score, each box numbered by its position:

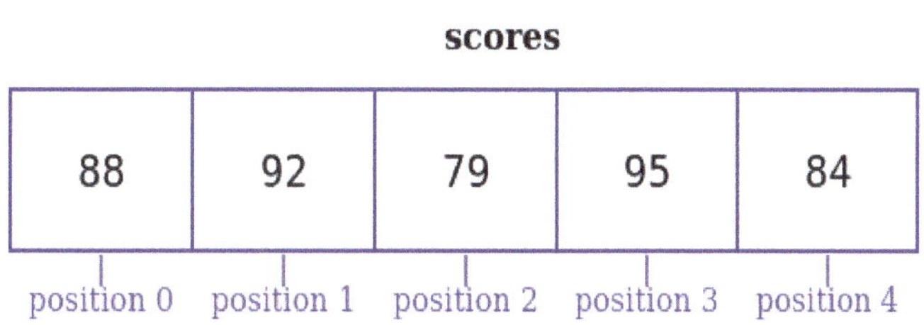

scores.length is 5 — but the last valid position is 4, not 5.

Notice the positions start at 0, not 1. This is a deliberate Java convention, not an arbitrary choice. The first box is position 0, the second is position 1, and so on. It feels unnatural at first, but it's consistent across the entire language, and you'll get used to it quickly.

With this picture in mind, computing the average becomes: set a running total to 0 (Chapter 2's pattern), then loop through every position from 0 to the last one (Chapter 4's pattern), adding each box's value to the total along the way. Nothing new is being invented here. This chapter is about combining two tools you already have with one new container to hold them.

6.2.1 Worked Example: From Blueprint to Pseudocode

Here's the labeled-boxes picture above, written as pseudocode, the same style introduced in Chapter 4:

```
SET scores TO [88, 92, 79, 95, 84]
SET total TO 0
FOR EACH position FROM 0 TO 4
    SET total TO total + scores[position]
END FOR
SET average TO total / 5
```

Compare this to the loop pseudocode from section 4.2.1: the start, check, and update are still there, just hidden inside FOR EACH, and the one new piece is scores[position], reaching into the container at a specific position to pull out a single value.

6.3 Syntax Bridge: Declaring and Looping Through an Array

Here's the labeled-boxes container, written as a real Java array, with a for loop computing the average exactly like the pseudocode above.

```
public class QuizAverage {
    public static void main(String[] args) {
        int[] scores = {88, 92, 79, 95, 84};
        int total = 0;

        for (int i = 0; i < scores.length; i = i + 1) {
            total = total + scores[i];
        }

        double average = total / 5.0;
        System.out.println(average);
    }
}
```

🔍 **SYNTAX GLOSSARY**

Decoding the Java on this page, piece by piece

`int[] scores = {88, 92, ...};` = *create an array called scores, a single container holding five int values in order*

`int[]` = *this variable holds a whole row of int values, not just one*

`scores.length` = *the number of boxes in the scores array, here, 5*

`scores[i]` = *reach into the array at position i and get that one value*

`for (int i = 0; i < scores.length; i = i + 1)` = *the exact loop pattern from Chapter 4, now visiting every position in the array*

Notice the loop condition is i < scores.length, not a hardcoded number like i < 5. This is deliberate: scores.length always reflects the array's real size, so this exact same loop works correctly whether scores holds 5 values or 5,000, without you changing a single character.

6.4 Predict, Observe, Explain (POE): Reaching Past the Last Box

Recall the off-by-one bug from Chapter 4's loop POE, the one caused by <= instead of <. Arrays have their own version of that exact same mistake, worth meeting on purpose here.

PREDICT

scores has 5 values, at positions 0 through 4. Suppose a loop's condition was accidentally written as i <= scores.length instead of i < scores.length. What do you predict happens when the loop tries to access scores[5]?

OBSERVE

There is no position 5. The valid positions are only 0, 1, 2, 3, and 4. Java doesn't quietly ignore this or return an empty value; it stops the program immediately with an error called ArrayIndexOutOfBoundsException. This is one of the most common errors beginning Java programmers see, and now you know exactly what causes it: asking for a box that doesn't exist.

EXPLAIN

In your own words, jot down: why does scores.length give you exactly one more than the array's last valid position? Connect this back to the boxes starting at position 0, not position 1.

This is exactly why the Syntax Bridge above used i < scores.length, not i <= scores.length. A one-character difference that's the line between a correct loop and a crashed program.

6.4.1 Worked Example: Changing a Value Inside an Array

Arrays aren't just for reading values back. You can update a single position directly, the same x = x + 1 update pattern from Chapter 2, just aimed at one box instead of a whole variable.

```
public class CurveScores {
    public static void main(String[] args) {
        int[] scores = {88, 92, 79, 95, 84};

        scores[2] = scores[2] + 5;

        for (int i = 0; i < scores.length; i = i + 1) {
            System.out.println(scores[i]);
        }
    }
}
```

🔍 **SYNTAX GLOSSARY**

Decoding the Java on this page, piece by piece

`scores[2] = scores[2] + 5;` *= take the value at position 2 (79), add 5, and store the result back at position 2*

Position 2 holds the third score (remember, positions start at 0), so this curves only the 79 up to 84, leaving every other position untouched. Trace through the printed output by hand: it should be 88, 92, 84, 95, 84, only the third line changed.

6.5 Practice Problems

Try tracing these by hand first, the way you traced the loop bugs in Chapter 4, then write the Java version using the patterns from this chapter.

- Given int[] temps = {68, 72, 65, 70, 74};, write a loop that prints every value in temps, one per line. Trace through it by hand first: what prints last?
- Write a loop that finds the highest value in temps from problem 1, without using any built-in shortcut methods. Just a loop and a variable that tracks the highest value seen so far.
- A loop is meant to print every value in an array of 8 elements, but its condition is written as i <= 8 instead of i < 8. Explain exactly what error this causes and why, connecting your answer to the POE in section 6.4.
- Refactor the Five Variables Problem from section 6.1 into a working Java array-and-loop solution that computes the average, the way section 6.3 did, but for seven scores instead of five.

ARCHITECTURAL AI AUDIT

The Scenario: you asked an AI assistant to write a method that finds the highest temperature in an array of daily readings. It produced the code below. The code is syntactically correct and will pass most quick tests, but it hides a real defect.

```
public static int findHighest(int[] temps) {
    int highest = temps[0];
    for (int i = 0; i <= temps.length; i = i + 1) {
        if (temps[i] > highest) {
            highest = temps[i];
        }
    }
    return highest;
}
```

> **TOOLING & WORKFLOW**
>
> **Augment:** For extra practice, generate your own version of this scenario using any AI assistant before comparing it to the one above.
>
> **Audit:** Do not ask AI to find the bug for you. That defeats the exercise.
>
> **Verify:** Use the compiler to confirm your fix actually resolves the problem.

13. Find the Defect: Run this method mentally against temps = {68, 72, 65, 70, 74}. Identify exactly which line is wrong, and name the specific exception it risks throwing.
14. Explain the Inconsistency: This bug will not crash on every array. Explain what has to be true about an array's contents for this bug to slip past casual testing undetected.
15. Correct the Code: Write a corrected version of findHighest.
16. Reflect: In one sentence, name the general principle this bug violates. You met this exact principle by name in section 6.4.

6.5.1 Companion Section: Two-Dimensional Arrays

Every array in this chapter so far held a single row of values: one list of temperatures, one list of scores. But some information is naturally organized as a grid, a seating chart, a spreadsheet, a game board. Java handles this with a two-dimensional array: an array where each position holds another array.

Think of it as a table with rows and columns. To reach any single value, you need two positions: which row, then which column.

```
int[][] grid = {
    {10, 20, 30},
    {40, 50, 60},
    {70, 80, 90}
};
```

🔍 **SYNTAX GLOSSARY**
Decoding the Java on this page, piece by piece

`int[][] grid` = *declares a 2D array of integers, two sets of brackets, one for rows and one for columns*

`grid[1][2]` = *reaches row 1, column 2 (0-indexed both ways), which holds 60 in the example above*

`grid.length` = *the number of rows*

`grid[0].length` = *the number of columns in the first row*

Looping through a 2D array uses two nested loops: an outer loop for rows, an inner loop for columns. This is exactly the nested loop pattern from Chapter 4, applied to a grid.

```
for (int row = 0; row < grid.length; row = row + 1) {
    for (int col = 0; col < grid[row].length; col = col + 1) {
        System.out.print(grid[row][col] + " ");
    }
    System.out.println();
}
```

This prints every row on its own line, moving left to right across each row before stepping down to the next. This is called row-major traversal, the standard pattern for most 2D array problems.

ONE RULE: before you move on

grid[row][col] always means row first, column second. Getting those backwards is the single most common 2D array mistake. Every time you write a 2D array access, say the indices aloud: row, then column.

A worked example: find the largest value in a 2D array of test scores. The logic is identical to finding the largest value in a 1D array from section 6.3, just with nested loops replacing the single loop.

```
int[][] scores = {
    {88, 92, 79},
    {95, 84, 91},
    {72, 88, 96}
};

int highest = scores[0][0];

for (int row = 0; row < scores.length; row = row + 1) {
    for (int col = 0; col < scores[row].length; col = col + 1) {
        if (scores[row][col] > highest) {
            highest = scores[row][col];
        }
    }
}

System.out.println("Highest score: " + highest);
```

Notice the only difference from the 1D version: two loop variables (row and col) instead of one (i), and two indices on the array access. The algorithm itself, track the highest seen so far, is unchanged.

- Trace through the 2D scores array above by hand and confirm the printed result. Then write a version that finds the lowest value instead.
- Write a nested loop that computes the sum of every value in a 2D array called data with 4 rows and 3 columns.

Optional: if you used an AI assistant on either of these, check its loop boundaries carefully. Off-by-one errors in nested loops are doubly common in 2D array code, since there are two conditions to get right instead of one.

KEY TAKEAWAYS

An array is a single named container holding many values of the same type, each reachable by its position.

Array positions start at 0, not 1. The last valid position is always one less than the array's length.

arrayName.length gives the number of boxes in the array; arrayName[i] reaches into position i to read or update a single value.

Looping through an array with i < arrayName.length is the standard, safe pattern. Using <= instead causes an ArrayIndexOutOfBoundsException.

Arrays only become useful in combination with loops from Chapter 4. Neither tool alone solves the Five Variables Problem.

6.7 Confidence Checkpoint

Quick gut-check before moving on. These aren't graded, they're just for you, to see what's already sticking.

CHECK YOURSELF

1. In your own words, what problem does an array solve that separate variables can't?
2. If an array has 10 values, what is the position number of the last one?
3. Why does i < scores.length avoid the bug that i <= scores.length causes?

If any of these feel shaky, that's useful information, not a failure. Flip back to that section before moving to Chapter 7.

Instructor Note: *This chapter intentionally limits itself to one-dimensional arrays of primitives (int), declared with a fixed initializer. Two-dimensional arrays, arrays of objects (revisited after Chapter 7), and dynamic resizing via ArrayList are deliberately deferred. Introducing a fixed-size, position-indexed container is enough conceptual weight for one Blueprint chapter. switch, and the loop control statements break and continue (flagged as deferred in Chapter 4's instructor note), are addressed in section 7.0, a short companion section at the start of the next chapter, now that arrays give break and continue a motivated, concrete use case (searching for a value and stopping early).*

CHAPTER 7

Class Design (OOP)

Building Your First Class

THE SIGNPOST: Where We're Headed

In this chapter, we will learn how to design and build our own object, deciding what it knows and what it can do, then writing that as a real Java class.

In Chapter 5, you learned to use objects someone else carefully designed, Scanner, String. You never needed to know how they worked inside, only how to call their methods. This chapter flips that completely: you're about to become the someone else.

7.0 Companion Section: switch, break, and continue

Before this chapter's main topic begins, here are three small but useful tools that build directly on conditionals (Chapter 3) and loops (Chapter 4), now that arrays (Chapter 6) give them a natural, motivating use case. None of this is new thinking, just new syntax for situations you've already recognized.

Recall the nested if/else-if chain from Chapter 3's door policy. When every branch is checking the exact same variable against a list of specific values, Java offers a more direct shape: the switch statement.

```
int dayNumber = 3;
switch (dayNumber) {
    case 1:
        System.out.println("Monday");
        break;
    case 2:
        System.out.println("Tuesday");
        break;
    case 3:
        System.out.println("Wednesday");
        break;
    default:
        System.out.println("Some other day");
}
```

🔍 SYNTAX GLOSSARY

Decoding the Java on this page, piece by piece

`switch (dayNumber) {` = *check dayNumber against a list of exact values, instead of writing if/else-if for each one*

`case 3:` = *if dayNumber equals 3, run the code starting here*

`break;` = *stop checking further cases once a match is found and handled. Without it, execution would fall through into the next case*

`default:` = *the catch-all, run if no case matched, the switch equivalent of a final else*

switch only makes sense when you're checking one variable against specific exact values. It cannot express a range check like age >= 18, the way Chapter 3's if/else can. For that reason, if/else remains the more general, more frequently used tool; switch is a convenience for the specific case of "one variable, many exact possibilities."

The keyword break above has a second, more common use: stopping a loop early. Recall Chapter 6's array search. Once a target value is found, there's no reason to keep checking the rest of the array.

```
int[] scores = {88, 92, 79, 95, 84};
int target = 79;

for (int i = 0; i < scores.length; i = i + 1) {
    if (scores[i] == target) {
        System.out.println("Found at position " + i);
        break;
    }
}
```

> 🔍 **SYNTAX GLOSSARY**
> *Decoding the Java on this page, piece by piece*
>
> **`break;`** = *immediately exits the loop entirely, skipping any remaining positions*

Without break, this loop would keep checking every remaining position even after finding a match, wasted work once the answer is already known. continue is break's sibling: instead of exiting the loop entirely, it skips just the current pass and moves on to the next one.

```
for (int i = 0; i < scores.length; i = i + 1) {
    if (scores[i] < 80) {
        continue;
    }
    System.out.println(scores[i]);
}
```

> 🔍 **SYNTAX GLOSSARY**
> *Decoding the Java on this page, piece by piece*
>
> **`continue;`** = *skip the rest of this pass and jump straight to the loop's next check, without exiting the loop*

This prints only the scores at or above 80, skipping the rest of that single pass's code without ending the loop early. Used carefully, break and continue make a loop's intent clearer, "stop once you've found it" or "skip this one and move on," instead of wrapping the loop's body in an extra layer of if/else just to express the same idea.

7.1 Discovery Challenge: The Student ID Card

Imagine you're designing a new student ID card system for a school. Before any software exists, someone has to decide: what information does every student ID need to hold, and what should the card system be able to do with that information?

On paper, sketch a simple index card for "Student." Split it into two halves: the top half lists what a Student needs to remember (information), and the bottom half lists what a Student should be able to do (actions).

ONE RULE: Read this before you start

Everything on your card must describe a Student in general, not one specific student. "Name" belongs on the card. "Ada Lovelace" does not. That's a specific value a real student would have, not a property every student needs. Keep your card describing the category, not one example of it.

TRY IT: Spend about 8 minutes on this

Fill out your Student card: at least three pieces of information every student needs, and at least two actions a student object should be able to perform (think about what a school's system might actually need to do with student data).

Most people land on information like: name, student ID number, grade level, GPA. And actions like: enroll in a class, check current GPA, print an ID badge. Notice the split again. The top half is always nouns (things the student has), the bottom half is always verbs (things the student does).

This is the core discovery of this chapter: every object you'll ever design splits cleanly into these same two categories, information it holds, and actions it can perform. Once you can sort a real-world idea into these two lists, you're most of the way to designing a class.

7.1.1 Worked Example: A Completed Student Card

This style of card, splitting an idea into what it knows and what it does, has a real name in software design: a CRC card, short for Class, Responsibility, Collaborator. Here's a filled-out version for Student:

Class: Student

Information it holds:

- name
- studentId
- gpa

Actions it can perform:

- introduce itself (print a greeting using its name)
- report its current GPA

Notice this card is deliberately small. A real student record system might track dozens of fields, but a first design should stay minimal. You can always add more later. Starting small and adding on is far easier than starting too big and untangling a mess.

7.2 Predict, Observe, Explain (POE): Information or Action?

Some items are trickier to sort than they first appear. Consider “full name” on a Student card, where the card already separately stores firstName and lastName.

PREDICT

Before reading further: should “full name” be stored as its own piece of information on the card, alongside firstName and lastName, or should it be an action the Student performs (something computed on demand, by combining firstName and lastName)?

OBSERVE

Most experienced designers choose the action: a getFullName() behavior that combines firstName and lastName every time it's asked, rather than storing a separate fullName field. Here's why it matters: if firstName ever changes (a legal name change, a typo correction), a stored fullName field could go stale and contradict the data it was built from. An action that recomputes it fresh every time can never be wrong in that way.

EXPLAIN

In your own words, jot down: why might storing the same fact in two different places on a class card (like a separate fullName alongside firstName and lastName) lead to bugs later, even if it seems convenient now?

This is a real design judgment call you'll keep making as you design more classes: when in doubt, prefer computing a value from existing information over storing it separately and risking the two falling out of sync.

7.3 Naming What You Already Did

FORMAL CONCEPT

A blueprint that describes what kind of information an object holds and what actions it can perform is called a class. The information slots are called fields, and the actions are called methods, the exact same word from Chapter 5, but now you're the one defining them, not just calling them.

Your Student card already was a class design, just written on paper instead of in Java. The top half (name, studentId, gpa) becomes fields. The bottom half (introduce itself, report GPA) becomes methods. Java just requires a precise grammar for writing both down.

7.3.1 A New Kind of Thinking Begins Here

Something shifted in this chapter that's worth naming directly. Chapters 1 through 6 were mostly about learning Java's grammar: how to write a correct instruction, a correct loop, a correct array. This chapter asked a different kind of question first: not "how do I write this?" but "what should this thing even be?", deciding what a Student needs to know and do, before any code existed at all.

That shift has a name: software design, the practice of deciding how a program's pieces should be organized and how they should relate to one another, separate from the syntax used to express those decisions. The CRC card you built in section 7.1 was a piece of software design. So was the judgment call in section 7.2, about whether fullName should be stored or computed.

From here through Chapter 14, this book asks both kinds of questions side by side: the syntax questions you're used to (how do I write private, extends, @Override), and the design questions underneath them (when should I make a field private, when does inheritance fit, when doesn't it). Both matter. Syntax without design produces code that runs but doesn't hold together. Design without syntax never becomes a real program at all.

7.4 Syntax Bridge: Writing Your First Class

Here's the Student card from section 7.1.1, written as a real Java class.

```
public class Student {
    String name;
    int studentId;
    double gpa;

    Student(String studentName, int id, double startingGpa) {
        name = studentName;
        studentId = id;
        gpa = startingGpa;
    }

    void introduce() {
        System.out.println("Hi, I'm " + name + ".");
    }

    void reportGpa() {
        System.out.println(name + "'s GPA is " + gpa);
    }
}
```

🔍 **SYNTAX GLOSSARY**
Decoding the Java on this page, piece by piece

`String name; int studentId; double gpa;` = *the fields, the "information it holds" half of the card, with no values yet*

`Student(String studentName, int id, double startingGpa) {` = *the constructor, runs once, when a new Student is created, to fill in the fields*

`void introduce() {` = *a method, one of the "actions it can perform" from the card. void means it doesn't hand back a result*

Notice the constructor's name is exactly Student, matching the class name. That's how Java recognizes it as the constructor, not just a regular method. This is also the exact pattern behind new Scanner(System.in) from Chapter 5: Scanner is a class, just like Student, and new Scanner(System.in) is calling Scanner's constructor, the same way you'll soon call Student's.

7.4.1 Worked Example: Creating and Using a Student Object

A class by itself doesn't do anything. It's a blueprint. To actually use it, you create an object from the class, the same new keyword from Chapter 5's new Scanner(System.in), just building a Student instead.

```
public class Main {
    public static void main(String[] args) {
        Student ada = new Student("Ada", 1001, 3.9);

        ada.introduce();
        ada.reportGpa();
    }
}
```

🔍 **SYNTAX GLOSSARY**
Decoding the Java on this page, piece by piece

`Student ada = new Student("Ada", 1001, 3.9);` = *create one specific Student object, named ada, with these three field values*

`ada.introduce();` = *call the introduce method on the ada object, the exact dot-notation from Chapter 5*

Notice the relationship: Student is the blueprint, the class. It describes what any student has and does, in general. ada is one specific object built from that blueprint, a real student with real values filled in. You could create a second student, maria, from the exact same Student class, with completely different field values, and both objects would work independently.

7.5 Practice Problems

Try the design step on paper first, the way you built the Student card, then write the Java version using the patterns from this chapter.

17. Design a CRC card for a class called Book, with at least three fields and two methods. Sort each item carefully into information versus actions, using the rule from section 7.1.
18. Translate your Book card from problem 1 into a real Java class, following the Student class in section 7.4 as a model: fields, a constructor, and at least two methods.
19. Write a short Main class that creates two different Book objects from your class in problem 2, with different field values, and calls a method on each one.
20. Look back at the POE in section 7.2. Design a Rectangle class with fields width and height. Should area be a stored field, or a method that computes width times height? Justify your answer using the same reasoning from the POE.

Optional: if you used an AI assistant to design any of these classes, sort its fields and methods into the CRC card categories yourself afterward, the way section 7.1 taught. An AI's class design is a starting point for your judgment, not a substitute for it.

7.6 Chapter Summary

KEY TAKEAWAYS

A class is a blueprint describing what information an object holds (fields) and what actions it can perform (methods).

A constructor runs once, when an object is created with new, to fill in that object's fields.

A class describes a category in general; an object built from new ClassName(...) is one specific instance with its own real values.

When a value can be computed from other fields, prefer a method that calculates it fresh over storing it separately. Stored duplicates can fall out of sync.

7.7 Confidence Checkpoint

Quick gut-check before moving on. These aren't graded, they're just for you, to see what's already sticking.

CHECK YOURSELF

1. In your own words, what's the difference between a class and an object?
2. What does a constructor do, and when does it run?
3. How is new Student("Ada", 1001, 3.9) similar to new Scanner(System.in) from Chapter 5?

If any of these feel shaky, that's useful information, not a failure. Flip back to that section before moving to Chapter 8.

Instructor Note: *Section 7.0 is a deliberate detour before this chapter's real topic begins. It resolves syntax promised but deferred in Chapter 3 (switch) and Chapter 4 (break, continue), now that Chapter 6's arrays give break and continue a concrete, motivating use case. It introduces no new thinking, only new syntax for situations students already recognize, and could be skipped or assigned as independent reading without disrupting the chapter's main Class Design arc. This chapter deliberately leaves every field public-by-default (no access modifiers), and the this keyword is avoided in favor of distinctly named constructor parameters (studentName vs. name) to sidestep shadowing confusion on a first encounter. Both of these are revisited in Chapter 8 (Encapsulation), which is explicitly framed as the "refinement pass" on the class built here. The same Student class is carried forward and tightened in that chapter, rather than starting over with a new example.*

CHAPTER 8

Encapsulation and Data Hiding (OOP)

Protecting Internal Data

THE SIGNPOST: Where We're Headed

In this chapter, we will learn how to protect an object's information from being set to nonsense values, by controlling exactly how the outside world is allowed to read and change it.

This chapter doesn't introduce a new class. It picks up the exact Student class you built in Chapter 7 and finds a real problem with it, then fixes that problem properly. This is what we called the “refinement pass” back when we planned this book: build first, protect second.

8.1 Discovery Challenge: Breaking Your Own Student Class

Here's the Student class exactly as you left it at the end of Chapter 7.

```
public class Student {
    String name;
    int studentId;
    double gpa;

    Student(String studentName, int id, double startingGpa) {
        name = studentName;
        studentId = id;
        gpa = startingGpa;
    }

    void reportGpa() {
        System.out.println(name + "'s GPA is " + gpa);
    }
}
```

Because every field is public by default, nothing stops code anywhere else in the program from reaching directly into a Student object and changing a field, completely bypassing the constructor.

```
Student ada = new Student("Ada", 1001, 3.9);
ada.gpa = -5.0;
ada.gpa = 400.0;
ada.studentId = -1;
```

ONE RULE: Read this before you start

All four lines above are completely legal Java. The compiler will not stop a single one of them. Your task is not to find a syntax error, because there isn't one. Your task is to explain, in plain English, what's wrong with each line anyway.

TRY IT: Spend about 5 minutes on this

For each of the four lines above, write one sentence explaining why that value doesn't make sense for a real student, even though Java allows it without complaint.

A GPA of -5.0 or 400.0 isn't a real GPA. A normal scale runs from 0.0 to 4.0. A negative studentId doesn't correspond to any real student record. Nothing about the Student class itself prevents any of this; the fields are wide open, and Java has no idea what counts as a "reasonable" GPA unless you tell it.

This is the core discovery of this chapter: a class with public fields offers no protection at all. Anyone using your Student class, including a future version of you, six months from now, half-asleep and rushing, can quietly corrupt an object's data with a single careless line, and Java will never complain.

8.2 Logical Blueprint: The Vending Machine, Revisited

Recall the vending machine from Chapter 5: you could use it completely through its buttons and slots, without ever opening it up and touching the wiring directly. That "no opening the machine" rule wasn't just a fun comparison. It's the exact protection this chapter is about to add to Student.

Before writing any code, let's plan the fix in plain English: instead of letting outside code touch ada.gpa directly, we want every read and every write to gpa to pass through a checkpoint, a method that can inspect the new value and reject it if it's nonsense, the same way a vending machine's coin slot can reject a fake coin before it ever reaches the machine's insides.

- Make the fields private, so they're no longer directly reachable from outside the class, like sealing the vending machine's case shut.
- Add a public method to read each field safely (a "getter"), like the machine's display screen showing you what's inside, without letting you reach in.

- Add a public method to update each field safely (a “setter”), which checks the new value before accepting it, like the coin slot rejecting anything that isn't a real coin.

Notice this plan doesn't remove any functionality. Outside code can still read and update a Student's GPA. It just forces those reads and writes through a checkpoint that can say no.

8.3 Syntax Bridge: private Fields, Getters, and Setters

Here's the Student class from Chapter 7, refined with the plan from section 8.2.

```
public class Student {
    private String name;
    private int studentId;
    private double gpa;

    Student(String studentName, int id, double startingGpa) {
        name = studentName;
        studentId = id;
        setGpa(startingGpa);
    }

    public double getGpa() {
        return gpa;
    }

    public void setGpa(double newGpa) {
        if (newGpa >= 0.0 && newGpa <= 4.0) {
            gpa = newGpa;
        } else {
            System.out.println("Invalid GPA, value not changed.");
        }
    }

    void reportGpa() {
        System.out.println(name + "'s GPA is " + gpa);
    }
}
```

🔍 **SYNTAX GLOSSARY**
Decoding the Java on this page, piece by piece

`private double gpa;` = *gpa can no longer be reached directly from outside this class, sealed, like the vending machine's case*

`public double getGpa() {` = *a getter, the public, safe way to read gpa from outside the class*

`public void setGpa(double newGpa) {` = *a setter, the public, safe way to update gpa, with a checkpoint built in*

`if (newGpa >= 0.0 && newGpa <= 4.0) {` = *the checkpoint itself, using Chapter 3's if/else and a new combined condition*

`newGpa >= 0.0 && newGpa <= 4.0` = *&& means "both conditions must be true," newGpa must be at least 0.0 AND at most 4.0*

Notice the constructor now calls setGpa(startingGpa) instead of writing gpa = startingGpa; directly. Even the constructor goes through the checkpoint. This means there is now exactly one place in the entire class where gpa actually gets written, and that one place always checks the value first.

8.4 Predict, Observe, Explain (POE): Testing the New Checkpoint

Let's revisit the exact four lines from the Discovery Challenge, now that gpa is private and protected by setGpa().

PREDICT

Before reading further: what happens now if code outside the class tries ada.gpa = -5.0; directly, the same line from section 8.1? And what happens if it instead calls ada.setGpa(-5.0);?

OBSERVE

ada.gpa = -5.0; no longer compiles at all. Java refuses to build the program, because gpa is private and unreachable from outside the class. This is a stronger protection than a runtime check: the bad line is caught before the program ever runs. Meanwhile, ada.setGpa(-5.0); does compile and run, but the checkpoint inside catches it: it prints "Invalid GPA, value not changed." and gpa keeps its previous, valid value.

EXPLAIN

In your own words, jot down: why is it meaningfully better for ada.gpa = -5.0; to fail to compile, rather than compiling and just silently producing a wrong GPA?

This is encapsulation's real payoff: bad data is caught as early as possible, ideally before the program even compiles, and at worst the moment it's attempted, rather than silently corrupting an object and causing a confusing bug somewhere else entirely, much later.

8.5 Refactor: Updating the Rest of the Program

Recall the Main class from Chapter 7's worked example, which called ada.reportGpa() directly. Now that gpa is private, let's revisit any code that used to touch fields directly and refactor it to use the new getter instead.

```
public class Main {
    public static void main(String[] args) {
        Student ada = new Student("Ada", 1001, 3.9);

        ada.reportGpa();
        System.out.println("Direct access: " + ada.getGpa());

        ada.setGpa(4.5);
        ada.setGpa(3.95);
    }
}
```

Notice ada.reportGpa() still works completely unchanged. It's a method defined inside the Student class itself, so it's allowed to read the private gpa field directly. Only code outside the class is blocked from touching gpa directly and must go through getGpa() and setGpa() instead. The first setGpa(4.5) call is silently rejected by the checkpoint (4.5 is above 4.0); the second, setGpa(3.95), succeeds.

8.6 Practice Problems

Try tracing these by hand first, the way you traced the door check in Chapter 3, then write the Java version using the patterns from this chapter.

21. Add a private field age to the Student class, with a getter getAge() and a setter setAge() that rejects any value below 0 or above 120. Explain your reasoning for the bounds you chose.
22. Trace through this sequence of calls by hand: Student m = new Student("Maria", 1002, 3.5); m.setGpa(-1.0); m.setGpa(2.8); System.out.println(m.getGpa());. What gets printed, and why?
23. Look back at the POE in section 8.4. Explain, in your own words, why making a field private and forcing all access through a setter is described as "encapsulation," connecting your answer to the vending machine analogy from Chapter 5 and section 8.2.
24. Design (on paper, CRC-card style) a private field and matching setter for a BankAccount class's balance, where the setter should reject any value that would make the balance negative. What should the setter do instead of

silently failing? Print a message, return a true/false success value, or something else? Justify your choice.

ARCHITECTURAL AI AUDIT

The Scenario: you asked an AI assistant to "add encapsulation" to a simple BankAccount class. It produced the code below. The code compiles and looks like a correct encapsulation example, but it has a real gap.

```
public class BankAccount {
    private double balance;
    public String accountHolderName;

    BankAccount(String name, double startingBalance) {
        accountHolderName = name;
        balance = startingBalance;
    }

    public double getBalance() {
        return balance;
    }

    public void setBalance(double newBalance) {
        if (newBalance >= 0) {
            balance = newBalance;
        }
    }
}
```

TOOLING & WORKFLOW

Augment: For extra practice, generate your own version of this scenario using any AI assistant before comparing it to the one above.

Audit: Do not ask AI to find the gap for you. That defeats the exercise.

Verify: Use the compiler to confirm your fix actually resolves the problem.

1. Find the Gap: Identify which field was never actually encapsulated, even though the class looks fully protected at a glance.
2. Explain the Risk: Describe a concrete, realistic scenario where that one unprotected field causes a real bug, even though balance itself stays perfectly safe.
3. Correct the Code: Write a corrected version of BankAccount.
4. Reflect: In one sentence, explain why "mostly encapsulated" offers a false sense of security compared to either fully public or fully private.

8.7 Chapter Summary

KEY TAKEAWAYS

Encapsulation means protecting an object's fields from being set to invalid values, by making them private and controlling access through methods.

A getter is a public method that safely returns a private field's value; a setter is a public method that safely updates it, often with a validity check.

Attempting to access a private field directly from outside its class is a compile-time error, caught before the program ever runs.

Code inside the class itself can still access its own private fields directly; the restriction only applies to code outside the class.

Routing even the constructor through a setter ensures there is exactly one place in the class where a field is ever actually written, and it's always checked.

8.8 Confidence Checkpoint

Quick gut-check before moving on. These aren't graded, they're just for you, to see what's already sticking.

CHECK YOURSELF

1. In your own words, what is encapsulation, and what problem does it solve?
2. What's the difference between a getter and a setter?
3. Why does ada.gpa = -5.0; fail to compile once gpa is private, while ada.setGpa(-5.0); compiles but does nothing harmful?

If any of these feel shaky, that's useful information, not a failure. Flip back to that section before moving to Chapter 9.

Instructor Note: *This chapter keeps name and studentId without setters that validate complex rules (e.g., uniqueness of studentId), to keep the validation logic in setGpa() the single clear teaching example. The this keyword is still avoided here, consistent with Chapter 7's note; constructor parameters remain distinctly named (studentName vs. name) rather than using this.name = name. Introducing this is deferred to Chapter 9 (Inheritance), where constructor chaining via super() creates a natural, motivated need for it.*

CHAPTER 9

Inheritance and Class Hierarchies (OOP)

Building Hierarchies of Objects

THE SIGNPOST: Where We're Headed

In this chapter, we will learn how to build a new class that automatically reuses everything an existing class already has, adding only what's genuinely different.

By the end of Chapter 8, your Student class was solid, encapsulated, validated, working. This chapter asks: what happens when you need a second, closely related kind of student? Do you start over from scratch?

9.1 Discovery Challenge: The Graduate Student Problem

Suppose the school now needs to track graduate students too. A GraduateStudent has a name, a studentId, and a gpa, everything a regular Student has, plus one more thing: an advisorName, the faculty member supervising their research.

Using only what you knew at the end of Chapter 8, the obvious move is to copy the entire Student class, rename it GraduateStudent, and add the new field.

```
public class GraduateStudent {
    private String name;
    private int studentId;
    private double gpa;
    private String advisorName;

    GraduateStudent(String studentName, int id, double startingGpa,
String advisor) {
        name = studentName;
        studentId = id;
        setGpa(startingGpa);
        advisorName = advisor;
    }

    public double getGpa() { return gpa; }

    public void setGpa(double newGpa) {
        if (newGpa >= 0.0 && newGpa <= 4.0) {
            gpa = newGpa;
        } else {
            System.out.println("Invalid GPA, value not changed.");
        }
    }

    void reportGpa() {
        System.out.println(name + "'s GPA is " + gpa);
    }
}
```

ONE RULE: Read this before you start

Imagine, a week later, you discover a bug in setGpa(). The upper bound should have been 4.33, not 4.0, to account for honors grades. You are not allowed to forget to fix this in even one of your student-like classes. List every single place in your program you would now need to find and apply this exact same fix.

TRY IT: Spend about 5 minutes on this

If GraduateStudent was built by copy-pasting Student, and a third class, ExchangeStudent, was later built the same way, write out exactly how many places the setGpa() bug fix from the One Rule above would need to be applied. How likely do you think it is that a real programmer, under deadline pressure, would actually remember every single one?

The honest answer is three separate places, in three separate files, and it's entirely plausible that a busy programmer fixes Student and GraduateStudent but completely forgets ExchangeStudent exists, because nothing in the code connects the three classes to each other. They look related only because a human remembers copy-pasting them. Java has no idea they're related at all.

This is the core discovery of this chapter: copy-pasting a class to create a related one duplicates not just today's code, but every future bug fix, forever, across every copy. What's needed instead is a way to say, formally, "GraduateStudent is a Student, plus a little more," so a fix to Student's shared logic only ever needs to happen once.

9.2 Logical Blueprint: "is-a" Relationships

Before writing any code, let's formalize what "GraduateStudent is a Student, plus a little more" actually means. A GraduateStudent should have every field and every method Student has, automatically, without retyping them, plus its own additional field, advisorName.

- Student is the general case: name, studentId, gpa, and the methods that go with them.
- GraduateStudent is a more specific case: everything Student has, plus advisorName.
- This relationship is called "is-a": a GraduateStudent is a Student. Compare this to a relationship that would NOT fit this pattern. A Course is not a Student, even though a Student takes Courses. "is-a" only applies when the more specific thing truly is a more specific version of the general thing.

Java has a built-in way to express exactly this "is-a" relationship, so that GraduateStudent can formally declare "I am a Student, plus more" and automatically gain every field and method Student already has, with zero copy-pasting.

9.3 Syntax Bridge: extends and super

Here's GraduateStudent rebuilt properly, using Java's extends keyword to formally declare the "is-a" relationship with Student, instead of copy-pasting Student's code.

```
public class GraduateStudent extends Student {
    private String advisorName;

    GraduateStudent(String studentName, int id, double startingGpa,
String advisor) {
        super(studentName, id, startingGpa);
        this.advisorName = advisor;
    }

    public String getAdvisorName() {
        return this.advisorName;
    }
}
```

🔍 **SYNTAX GLOSSARY**

Decoding the Java on this page, piece by piece

`public class GraduateStudent extends Student {` = *declares the is-a relationship: GraduateStudent automatically gains every field and method from Student*

`super(studentName, id, startingGpa);` = *calls Student's own constructor first, to set up name, studentId, and gpa exactly like Chapter 8 already does, no copy-pasting that logic*

`this.advisorName = advisor;` = *this refers to "the object being built right now," needed here because the parameter and field share a similar role*

`this.advisorName` = *the period after this works exactly like the dot-notation from Chapter 5, reaching into this specific object's own field*

Notice GraduateStudent's class body is now tiny, just the one new field and a constructor. getGpa(), setGpa(), and reportGpa() are not retyped anywhere; they're inherited automatically from Student. A GraduateStudent object can call grad.getGpa() and it works, even though getGpa() doesn't appear anywhere in the GraduateStudent class itself.

Also notice super(...) must be the very first line inside the constructor. Java requires the general case (Student) to finish setting itself up before the more specific case (GraduateStudent) adds anything on top.

9.4 Predict, Observe, Explain (POE): Fixing the Bug, For Real This Time

Let's return to the setGpa() bug from the Discovery Challenge, the upper bound should be 4.33, not 4.0. This time, GraduateStudent was built with extends Student instead of copy-pasting.

PREDICT

Before reading further: if you fix the bound inside setGpa() in the Student class only, one single edit, in one single file, does a GraduateStudent object automatically benefit from that fix too, or does GraduateStudent still have the old, buggy bound?

OBSERVE

GraduateStudent automatically benefits from the fix, with zero additional edits. Since GraduateStudent doesn't have its own copy of setGpa(), it inherits Student's version entirely, there's only one setGpa() in the whole program to fix. This is the direct payoff of the Logical Blueprint from section 9.2: the one-edit, everywhere-fixed outcome that copy-pasting could never give you.

EXPLAIN

In your own words, jot down: connect this back to the Discovery Challenge. Why does extends eliminate the exact risk you identified there (forgetting to apply the same fix to every copy-pasted class)?

9.4.1 Worked Example: Student and GraduateStudent Together

Here's both classes in use side by side, showing exactly which methods come from where.

```
public class Main {
    public static void main(String[] args) {
        Student maria = new Student("Maria", 1002, 3.6);
        GraduateStudent ada = new GraduateStudent("Ada", 1001, 3.9,
"Dr. Turing");

        maria.reportGpa();
        ada.reportGpa();

        System.out.println(ada.getAdvisorName());
    }
}
```

🔍 **SYNTAX GLOSSARY**
Decoding the Java on this page, piece by piece

`ada.reportGpa();` = *GraduateStudent doesn't define reportGpa() itself, this call uses the inherited version from Student*

`ada.getAdvisorName();` = *this method only exists on GraduateStudent; maria.getAdvisorName() would not compile, since Student never had an advisor*

Notice maria, a plain Student, has no advisorName and no getAdvisorName() method. The “is-a” relationship only flows one direction. Every GraduateStudent is a Student, but not every Student is a GraduateStudent.

9.5 Practice Problems

Try the design step on paper first, the way you sorted is-a relationships in section 9.2, then write the Java version using the patterns from this chapter.

25. Decide, with a one-sentence justification: does a SavingsAccount “is-a” BankAccount? Does a Car “is-a” Engine? Explain why one fits the is-a pattern from section 9.2 and the other doesn't.
26. Design a class ExchangeStudent that extends Student, adding one new field, homeCountry, with a constructor and a getter. Follow the GraduateStudent class in section 9.3 as a model.
27. Trace through this by hand: GraduateStudent g = new GraduateStudent("Sam", 1003, 5.0, "Dr. Lovelace");. What gets printed when g.reportGpa() is called, and why? Connect your answer to setGpa()'s validation from Chapter 8.
28. Look back at the POE in section 9.4. Explain, in your own words, why a bug fix to a parent class's method automatically applies to every child class, without those child classes needing any changes at all.

ARCHITECTURAL AI AUDIT

The Scenario: you asked an AI assistant to create a Course class for a school system that already has the Student class from this chapter. It produced the code below. The code compiles and runs without errors, but it represents a real design liability.

```
public class Course extends Student {
    private String courseCode;
    private int maxEnrollment;

    Course(String courseName, int id, double credits, String code, int
max) {
        super(courseName, id, credits);
        courseCode = code;
        maxEnrollment = max;
    }

    public String getCourseCode() {
        return courseCode;
    }
}
```

TOOLING & WORKFLOW

Augment: For extra practice, generate your own version of this scenario using any AI assistant before comparing it to the one above.

Audit: Do not ask AI to find the flaw for you. That defeats the exercise.

Verify: Use the compiler to confirm your refactor still produces working code.

5. The Design Flaw: Using the is-a test from section 9.2, explain exactly why extends Student is the wrong design here, even though the code compiles and runs.
6. The Fragility Risk: Identify what a Course object now incorrectly inherits or exposes (think about what getGpa() or reportGpa() would mean, called on a Course). Explain why this becomes a real problem as the school system grows.
7. The Refactor: Propose a corrected design. It does not need to use inheritance at all.
8. Reflect: In one sentence, state the difference between classes that merely share fields and classes that share a true is-a relationship.

9.6 Chapter Summary

KEY TAKEAWAYS

Inheritance lets one class (the child) automatically reuse another class's (the parent's) fields and methods, using extends.

Inheritance should only be used for a true “is-a” relationship. A GraduateStudent is a Student, but a Course is not a Student.

super(...) calls the parent class's constructor and must be the first line in the child's constructor.

this refers to the specific object currently being built or used, distinguishing a field from a same-named parameter.

A fix to a parent class's method automatically applies to every child class. This is the core advantage inheritance has over copy-pasting.

9.7 Confidence Checkpoint

Quick gut-check before moving on. These aren't graded, they're just for you, to see what's already sticking.

CHECK YOURSELF

1. In your own words, what does it mean for one class to “extend” another?
2. What does super(...) do, and why must it be the first line of a child class's constructor?
3. Why is copy-pasting a class to create a related one risky, even if it works correctly on day one?

If any of these feel shaky, that's useful information, not a failure. Flip back to that section before moving to Chapter 10.

Instructor Note: *This chapter deliberately avoids method overriding (redefining a parent method inside a child class). GraduateStudent only adds new members, it never redefines an inherited one. Overriding is intentionally reserved for Chapter 10 (Polymorphism), where @Override has a clear motivating use case: making reportGpa() behave differently for a GraduateStudent while still being called the exact same way. Introducing both extends and overriding in the same chapter risks conflating “reuse” with “redefine,” which are different ideas students should meet separately.*

CHAPTER 10

Polymorphism (OOP)

Writing Flexible Interfaces

THE SIGNPOST: Where We're Headed
In this chapter, we will learn how to call the exact same method name on different kinds of objects, and have each one respond in its own appropriate way automatically.

Chapter 9 ended with GraduateStudent quietly reusing Student's exact reportGpa() method, unchanged. This chapter asks the opposite question: what if you want GraduateStudent to report its GPA a little differently, mentioning its advisor, say, without breaking the one-method-everywhere convenience inheritance just gave you?

10.1 Discovery Challenge: The Mixed Roster

Suppose the school wants to print a GPA report for an entire roster, a mix of regular Student objects and GraduateStudent objects together, in one list, in any order. Using only what you knew at the end of Chapter 9, you might try writing a loop that checks each object's type and calls something different depending on what it finds.

```
for (Student s : roster) {
    if (s instanceof GraduateStudent) {
        GraduateStudent g = (GraduateStudent) s;
        System.out.println(g.getName() + "'s GPA is " + g.getGpa()
            + ", advised by " + g.getAdvisorName());
    } else {
        System.out.println(s.getName() + "'s GPA is " + s.getGpa());
    }
}
```

ONE RULE: Read this before you start

Now imagine a third kind of student is added later, ExchangeStudent, from Chapter 9's practice problems, which should also print its home country. You are not allowed to forget to update this loop. Think for a moment about what would happen to this exact loop if you simply forgot.

TRY IT: Spend about 5 minutes on this

Write out, in plain English, every single place in a real program that would need a new else if branch added every time a new kind of Student is introduced. Then consider: who is responsible for remembering to make that update? The person adding the new student type, or the person who originally wrote this loop, possibly months or years earlier?

If a third or fourth kind of student is added and this loop isn't updated, ExchangeStudent quietly falls into the else branch and gets treated like a plain Student. No compiler error, no warning, just silently incomplete output. This should feel familiar: it's the exact same risk as Chapter 9's copy-pasted classes, just relocated from the class definitions into every loop that uses them.

This is the core discovery of this chapter: checking an object's specific type and branching by hand doesn't scale, for the same reason copy-pasting didn't scale in Chapter 9. What's needed is a way for each object to know how to report itself correctly, so the loop never needs to ask "what kind of student is this?" at all.

10.2 Logical Blueprint: Let the Object Decide

Before writing any code, let's flip the responsibility from section 10.1's loop. Instead of the loop asking "what kind of student is this, and what should I print?", each class should answer for itself: "Here's how I report my own GPA."

- Student already has a reportGpa() method. This becomes the default, general-case behavior.
- GraduateStudent should provide its own version of reportGpa(), one that also mentions the advisor, replacing the inherited one specifically for GraduateStudent objects.
- The loop itself should stay completely unchanged: call s.reportGpa() on every object in the roster, regardless of its specific type, and trust each object to do the right thing.

This shifts the responsibility for "what does reporting look like for me" from the loop (which has to know about every possible type) to each individual class (which only has to know about itself). Adding a new kind of student later means writing one new method in one new class. The loop is never touched again.

10.3 Syntax Bridge: @Override

Here's GraduateStudent, extended from Chapter 9, with its own version of reportGpa() that overrides, replaces, the one it would otherwise inherit from Student.

```
public class GraduateStudent extends Student {
    private String advisorName;

    GraduateStudent(String studentName, int id, double startingGpa,
String advisor) {
        super(studentName, id, startingGpa);
        this.advisorName = advisor;
    }

    public String getAdvisorName() {
        return this.advisorName;
    }

    @Override
    void reportGpa() {
        System.out.println(getName() + "'s GPA is " + getGpa()
            + ", advised by " + advisorName);
    }
}
```

🔍 SYNTAX GLOSSARY

Decoding the Java on this page, piece by piece

`@Override` = *tells Java (and any human reading the code) that this method is intentionally replacing an inherited one with the exact same name*

`void reportGpa() {` = *must match Student's reportGpa() exactly in name and parameters, this is what makes it an override, not a brand-new, unrelated method*

Notice @Override isn't strictly required for this to work. Java will override the method even without it. It's there as a safety net: if you misspell reportGpa() as reportGPA() by accident, @Override causes a compiler error immediately, instead of silently creating a second, unrelated method that never gets called. Always include it when you intend to override.

ONE RULE: professional practice

Always write @Override when overriding a method. This is not optional in professional Java. Every major style guide (Oracle, Google) and every code review culture requires it. The reason is not syntax, it is safety: @Override turns a silent, hard-to-find bug (wrong method name, wrong signature) into an immediate compiler error. In a production codebase, silent bugs cost hours. Compiler errors cost seconds.

10.4 Predict, Observe, Explain (POE): One Loop, Many Behaviors

Let's return to the Discovery Challenge's roster loop, simplified back down to its original, type-checking-free form: for (Student s : roster) { s.reportGpa(); }

PREDICT

Before reading further: if roster contains a mix of Student and GraduateStudent objects, and this simplified loop just calls s.reportGpa() on each one with no instanceof check at all, does each object still print correctly? A plain Student printing the simple message, and a GraduateStudent printing the advisor too?

OBSERVE

Yes, each object prints exactly the right message for its own actual type, with no type-checking code anywhere in the loop. When s.reportGpa() runs, Java looks at the real, actual type of the object s is currently pointing to, not just the Student label on the loop variable, and calls that object's own version of the method. A plain Student uses Student's version; a GraduateStudent uses its own overridden version. This is decided automatically, every single time, no matter how many kinds of students exist.

EXPLAIN

In your own words, jot down: connect this back to the Discovery Challenge. Why does this loop never need an else if branch added, no matter how many new kinds of Student get added in the future?

This is the entire idea behind the word polymorphism, from Greek roots meaning “many forms.” The same method call takes a different, appropriate form depending on which object it's actually called on.

10.5 Refactor: Deleting the Type-Checking Loop

Here's the full before-and-after. First, the original Discovery Challenge loop from section 10.1:

```
for (Student s : roster) {
    if (s instanceof GraduateStudent) {
        GraduateStudent g = (GraduateStudent) s;
        System.out.println(g.getName() + "'s GPA is " + g.getGpa()
            + ", advised by " + g.getAdvisorName());
    } else {
        System.out.println(s.getName() + "'s GPA is " + s.getGpa());
    }
}
```

And here's the refactored version, now that reportGpa() is overridden inside each class instead of branched on inside the loop:

```
for (Student s : roster) {
    s.reportGpa();
}
```

Both versions produce identical output. But the refactored version has zero knowledge of GraduateStudent, ExchangeStudent, or any future student type. It only knows about Student, and trusts polymorphism to handle the rest. This is the real payoff of this chapter: code that doesn't need to change when your class hierarchy grows.

10.5.1 The Fourth Pillar, Named

Object-oriented programming is traditionally organized around four pillars. You've now built all four, three of them across full chapters of their own, and one of them quietly, in nearly everything you've written since Chapter 5, without it ever being named.

- Encapsulation (Chapter 8): protecting an object's internal data, exposing only safe, controlled access to it.
- Inheritance (Chapter 9): building a more specific class that reuses a general class's fields and methods automatically.
- Polymorphism (this chapter): calling the same method name on different objects and getting correct, type-appropriate behavior automatically.
- Abstraction: exposing only what a user of a class needs to know, and hiding everything else. This is the one pillar that never got its own chapter, because you'd already been practicing it since Chapter 5: calling keyboard.nextLine() without knowing how Scanner reads keystrokes, or calling ada.getGpa() without knowing gpa is even stored as a field. Every getter, every method signature, every class you've used as a “black box” was abstraction at work.

Notice that abstraction isn't really a separate technique from the other three. It's closer to the reason the other three matter. Encapsulation is abstraction applied to an object's data. A well-designed method call is abstraction applied to behavior.

You weren't missing a fifth tool; you were using this one the whole time, under a name you hadn't met yet.

10.6 Practice Problems

Try the design step on paper first, the way you reasoned about the roster loop in section 10.2, then write the Java version using the patterns from this chapter.

29. Add an @Override of reportGpa() to the ExchangeStudent class you designed in Chapter 9's practice problems, so it also mentions homeCountry. Follow the GraduateStudent override in section 10.3 as a model.
30. Trace through this by hand: a roster array contains one Student and one GraduateStudent. A loop calls s.reportGpa() on each. Write out exactly what prints for each object, and explain why Java picks a different version of the method for each one.
31. Explain, in your own words, what would happen if you wrote void reportGPA() (different capitalization) in GraduateStudent with an @Override annotation above it, intending to override Student's reportGpa(). Would this compile? Why does @Override matter here?
32. Look back at the Discovery Challenge's instanceof loop in section 10.1. Explain, in your own words, why that loop has the same underlying problem as the copy-pasted classes from Chapter 9, even though no code was literally copy-pasted this time.

ARCHITECTURAL AI AUDIT

The Scenario: you asked an AI assistant to print a roster of mixed Student and GraduateStudent objects, after both classes already had a correctly working @Override of reportGpa(). It produced the code below. The code runs correctly, but it represents a real design liability.

```
for (Student s : roster) {
    if (s instanceof GraduateStudent) {
        s.reportGpa();
    } else {
        s.reportGpa();
    }
}
```

TOOLING & WORKFLOW

Augment: For extra practice, generate your own version of this scenario using any AI assistant before comparing it to the one above.

Audit: Do not ask AI to find the flaw for you. That defeats the exercise.

Verify: Use the compiler to confirm your refactor still produces identical output.

9. The Design Flaw: Trace both branches of this if/else by hand. What is the instanceof check actually accomplishing here?
10. The Fragility Risk: Explain why this pattern becomes a real liability as more student types are added in the future, connecting your answer to Chapter 9's Discovery Challenge about the setGpa() bug fix.
11. The Refactor: Rewrite this loop to remove the unnecessary instanceof check entirely.
12. Reflect: In one sentence, explain why a future developer reading the original code might wrongly assume the two branches need to behave differently, when they don't.

10.7 Chapter Summary

KEY TAKEAWAYS

Polymorphism means calling the same method name on different objects and having each one respond with its own appropriate behavior, automatically.

@Override lets a child class replace an inherited method with its own version, matching the original method's name and parameters exactly.

@Override isn't required to make overriding work, but it catches mistakes (like typos) as a compiler error instead of a silent bug.

Polymorphism replaces manual instanceof type-checking, which has the same scaling problem as copy-pasted classes: every new type requires hunting down and updating every check.

Code written against the general type (Student) automatically works correctly with every more specific type (GraduateStudent, ExchangeStudent, and any future type), with no changes required.

10.8 Confidence Checkpoint

Quick gut-check before moving on. These aren't graded, they're just for you, to see what's already sticking.

CHECK YOURSELF

1. In your own words, what is polymorphism?

2. What does @Override do, and what happens if you misspell the method name underneath it?

3. Why does manually checking instanceof in a loop have the same long-term problem as copy-pasting a class?

If any of these feel shaky, that's useful information, not a failure. Flip back to that section before moving to Chapter 11.

Instructor Note: *This chapter intentionally stops short of abstract classes and interfaces, which formalize abstraction and polymorphism further by requiring every subclass to provide its own version of a method. Student's reportGpa() here is a normal, fully-implemented method that happens to be overridden, not an abstract one. Section 10.5.1 names abstraction as a concept students have already practiced informally since Chapter 5, but it deliberately stops at that conceptual naming. It does not introduce the abstract keyword or the interface keyword, which remain out of scope and are left for a later course or appendix.*

CHAPTER 11
Recursion

Solving Problems by Self-Reference

THE SIGNPOST: Where We're Headed

In this chapter, we will learn how a method can solve a problem by calling a smaller version of itself, instead of using a loop.

Every repeating task you've written so far used a loop from Chapter 4. This chapter introduces a second, very different way to repeat: having a method call itself. It's one of the harder ideas in this book. If it doesn't fully click on the first pass, that's completely normal. We'll come back to it again in Chapter 12 with concrete, traceable examples.

11.1 Discovery Challenge: The Nesting Dolls

Picture a Russian nesting doll, a wooden doll that opens up to reveal a smaller doll inside, which opens to reveal an even smaller doll inside that, and so on, until you reach one tiny solid doll that doesn't open at all.

Suppose someone asks you: “How many dolls are in this set?” Without taking the whole set apart and counting all at once, describe a procedure for counting them, one doll at a time, using only this idea: “the count of dolls inside this doll equals 1 (for this doll itself) plus the count of dolls inside the next doll in.”

ONE RULE: Read this before you start

Your procedure must work by only ever looking at one doll at a time. The current one and the one immediately inside it. You are not allowed to look ahead and count the whole stack visually. And your procedure must say exactly what happens when you reach the tiny solid doll that doesn't open. Without that, your counting procedure never finishes.

TRY IT: Spend about 5 minutes on this

Write out your counting procedure as a short numbered list of plain-English steps. Make sure it explains what to do at a normal doll, and separately, what to do at the final, solid doll.

Most people land on something like: "If this doll doesn't open, the count is 1. Otherwise, the count is 1 plus whatever the count is for the doll inside it." Notice the strange shape of that second sentence. It defines "count" partly in terms of "count," just applied to a smaller doll.

This is the core discovery of this chapter: some problems are most naturally described in terms of a smaller version of the exact same problem, plus a little extra work, stopping only when the problem becomes too small to shrink further. That's the entire idea behind recursion. You've just designed your first recursive procedure, in plain English, with no code at all.

11.1.1 Worked Example: Counting Down, Two Ways

Here's a simpler, very common example of the same idea: counting down from a number to zero. You already know how to do this with a loop from Chapter 4:

```
public static void countDownLoop(int n) {
    for (int i = n; i >= 0; i = i - 1) {
        System.out.println(i);
    }
}
```

Here's the exact same behavior, described recursively instead, using the nesting-doll instinct from section 11.1: "print n, then count down from one smaller than n, unless n is already 0, in which case just print 0 and stop."

```
public static void countDownRecursive(int n) {
    System.out.println(n);
    if (n > 0) {
        countDownRecursive(n - 1);
    }
}
```

Notice countDownRecursive calls itself, with a smaller input (n - 1) each time, exactly like opening one doll to find a smaller version of the same doll inside. And notice the if (n > 0) check: this is the "solid doll that doesn't open" from section 11.1, the point where the method stops calling itself.

11.2 Naming What You Already Did

FORMAL CONCEPT

A method that solves a problem by calling itself with a smaller version of the same problem is called recursive. The smallest version of the problem, which the method can answer directly without calling itself again, is called the base case. Every other call, which does a little work and then calls itself again with a smaller input, is called the recursive case.

In countDownRecursive, the base case is n == 0 (print 0, then stop, no further call). The recursive case is everything else (print n, then call countDownRecursive(n - 1)). Every recursive method you'll ever write needs both pieces: at least one base case to stop the chain, and at least one recursive case that makes genuine progress toward that base case.

11.3 Predict, Observe, Explain (POE): The Method Without a Floor

Recall the infinite loop from Chapter 4, caused by a loop with no way to ever make its condition false. Recursion has its own version of that same failure, worth meeting on purpose here.

PREDICT

Suppose the if (n > 0) check was accidentally left out of countDownRecursive entirely, so the method always calls itself again no matter what n becomes. What do you predict happens when countDownRecursive(3) is called?

OBSERVE

The method keeps calling itself forever, 3, 2, 1, 0, -1, -2, and on without end, since nothing ever stops it. But unlike Chapter 4's infinite loop, which can run forever without crashing, a runaway recursive call usually crashes the program quickly, with an error called a StackOverflowError. Each call waits on the one before it to finish, and Java can only keep track of so many waiting calls at once before running out of space to track them.

EXPLAIN

In your own words, jot down: connect this back to the nesting dolls in section 11.1. What would a nesting doll set be like if it had no smallest, solid doll? Why does that make the dolls impossible to fully count?

A missing or incorrect base case is the recursive equivalent of Chapter 4's missing update line, both remove the one thing that was supposed to eventually stop the repetition.

11.4 Call Stack Blueprint: Tracing countDownRecursive(3)

To really see what's happening inside a recursive call, it helps to draw it out as a stack of waiting calls, growing downward as each call makes its own call, then shrinking back upward as each one finishes.

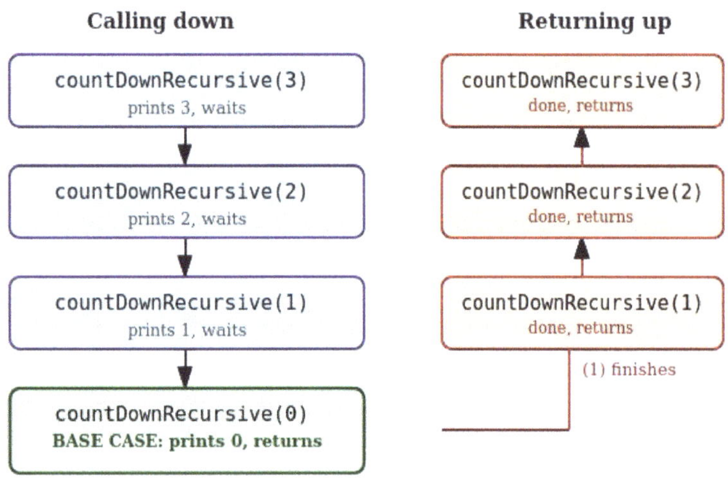

Each call waits for the one below it to finish completely before it can finish and return itself.

Each box represents one call, waiting on the call below it to completely finish before it can finish itself. countDownRecursive(3) prints 3, then calls countDownRecursive(2) and waits. countDownRecursive(2) prints 2, then calls countDownRecursive(1) and waits. This continues until countDownRecursive(0), the base case, prints 0 and returns immediately, with no further call. Only then does the stack unwind: countDownRecursive(1) finishes, then countDownRecursive(2), then finally countDownRecursive(3).

Notice this is exactly why a missing base case causes a crash: with no box ever finishing, the stack only ever grows downward, call after call, until Java runs out of room to track them all.

11.5 Practice Problems

Try tracing these by hand first, the way you traced countDownRecursive(3) in section 11.4, draw your own stack of boxes before writing or running any Java.

- Write a recursive method countUp(int n, int max) that prints every number from n up to max, then stops. Identify its base case and its recursive case before writing any code.
- Draw the call stack, the way section 11.4 did, for countDownRecursive(2). How many boxes does your diagram have in total, including the base case?

- A recursive method factorial(int n) is supposed to return 1 when n is 0, and otherwise return n times factorial(n - 1). Identify the base case and the recursive case in this plain-English description, the same way section 11.2 identified them for countDownRecursive.
- Look back at the POE in section 11.3. Explain, in your own words, why a StackOverflowError and an infinite loop are both caused by a missing stopping condition, but produce different symptoms (a crash versus a program that simply never finishes).

Optional: if you used an AI assistant to write a recursive method here, find its base case yourself before trusting it works, the same identification step section 11.2 walked through. A missing or wrong base case can look fine in an AI's explanation and still crash the moment it runs.

11.6 Chapter Summary

KEY TAKEAWAYS

A recursive method solves a problem by calling itself with a smaller version of the same problem.

Every recursive method needs a base case (a smallest version it can answer directly) and a recursive case (work plus a smaller call).

A missing or unreachable base case causes a StackOverflowError, the recursive equivalent of an infinite loop.

A call stack diagram traces recursive calls as boxes that grow downward (calling) and then unwind upward (returning), one finished call at a time.

11.7 Confidence Checkpoint

Quick gut-check before moving on. These aren't graded, they're just for you, to see what's already sticking.

CHECK YOURSELF

1. In your own words, what is a base case, and why does every recursive method need one?

2. In countDownRecursive, what makes n - 1 a "smaller version of the same problem" rather than just a different problem entirely?

3. What causes a StackOverflowError, and how is it similar to an infinite loop?

If any of these feel shaky, that's genuinely expected for a first encounter with recursion, not a failure. We'll return to it with more worked examples in Chapter 12, including recursive search and sort, traced the same way as this chapter.

Instructor Note: *This chapter is intentionally the lighter "Touch 1" pass on recursion, per the book's two-touch design for conceptually heavy topics. It covers only void, non-returning recursive methods (countDownRecursive) so students can focus entirely on the call-and-return shape without also tracking a returned value. Recursive methods that return values (like factorial, hinted at in Practice Problem 3, and recursive binary search) are deliberately deferred to Chapter 12, where they are paired directly against their iterative equivalents from Chapters 4 and 6 for direct comparison. This is the "Touch 2" Blueprint-level treatment.*

CHAPTER 12

Algorithm Analysis

Searching and Sorting, Iteratively and Recursively

THE SIGNPOST: Where We're Headed

In this chapter, we will learn standard ways to find a value inside a list and put a list in order, and we'll see each one written both with a loop and with recursion, side by side.

This chapter brings together more tools from earlier in the book than any other so far: arrays from Chapter 6, loops from Chapter 4, and recursion from Chapter 11. It's also where recursion gets its second, deeper pass. You'll see it solve real problems this time, not just count down.

12.1 Discovery Challenge: Finding a Name in a Phone Book

Suppose you're looking for "Martinez" in a printed, alphabetically sorted phone book with a thousand pages. Describe, in plain English, two different ways you might search for it: one where you start at page 1 and check every single page in order, and one that takes advantage of the fact that the book is alphabetically sorted.

ONE RULE: Read this before you start

For your second method, you are not allowed to flip through pages one at a time. You must describe a method that uses the book's alphabetical order to skip large sections entirely. Think about how you'd actually use a phone book in real life, opening to a guessed spot rather than starting from page 1.

TRY IT: Spend about 5 minutes on this

Write both methods as short numbered lists. Then estimate: for a 1,000-page book, roughly how many pages would each method check in the worst case before finding "Martinez" or confirming it isn't there?

The first method, checking page 1, then 2, then 3, and so on, could need up to 1,000 checks in the worst case. The second method is closer to: open to the middle page. If "Martinez" would come before what's on that page, the entire second half of the book can be eliminated at once; if it would come after, the entire first half can be eliminated. Repeating this, each check eliminates roughly half of what's left, reaching a single page in around 10 checks instead of 1,000.

This is the core discovery of this chapter: the same task, finding something, can be solved with wildly different amounts of work depending on the strategy, even on the exact same data. A sorted list isn't just nicer to read; it unlocks a fundamentally faster way to search it.

12.2 Syntax Bridge: Linear Search

Here's the first method, checking every position in order, written as a real Java method, using the array and loop patterns from Chapters 4 and 6.

```java
public static int linearSearch(int[] numbers, int target) {
    for (int i = 0; i < numbers.length; i = i + 1) {
        if (numbers[i] == target) {
            return i;
        }
    }
    return -1;
}
```

🔍 **SYNTAX GLOSSARY**

Decoding the Java on this page, piece by piece

`return i;` = *immediately exits the method, handing back the position where target was found*

`return -1;` = *if the loop finishes without finding target anywhere, -1 signals "not found," a position that could never be real*

This is exactly your "check every page" method from section 12.1, just written in Java. In the worst case, the target is the very last element, or isn't present at all, this checks every single element, just like checking all 1,000 phone book pages.

12.3 Syntax Bridge: Binary Search (Iterative)

Here's your phone-book "skip half" method, written iteratively. This only works correctly on a sorted array. That's the one requirement that unlocks the speed.

```
public static int binarySearchIterative(int[] numbers, int target) {
    int low = 0;
    int high = numbers.length - 1;

    while (low <= high) {
        int mid = (low + high) / 2;
        if (numbers[mid] == target) {
            return mid;
        } else if (numbers[mid] < target) {
            low = mid + 1;
        } else {
            high = mid - 1;
        }
    }
    return -1;
}
```

🔍 **SYNTAX GLOSSARY**

Decoding the Java on this page, piece by piece

`int low = 0; int high = numbers.length - 1;` = *the boundaries of the section still worth checking, starts as the whole array*

`int mid = (low + high) / 2;` = *the middle position of the current section, the "open to the middle page" move from section 12.1*

`low = mid + 1;` = *target is bigger than the middle value, so eliminate the entire left half*

`high = mid - 1;` = *target is smaller than the middle value, so eliminate the entire right half*

Each pass through the loop eliminates roughly half of what's left, exactly like the phone book. Trace through binarySearchIterative on a sorted array of 8 elements by hand: notice it takes at most 3 checks to find anything, where linearSearch could take up to 8.

12.4 Touch 2: Binary Search, Recursively

Chapter 11 introduced recursion using countDownRecursive, a method that didn't return anything, just printed and stopped. This is where recursion does real, useful work: binary search, solved by calling a smaller version of the exact same search.

```
public static int binarySearchRecursive(int[] numbers, int target, int
low, int high) {
    if (low > high) {
        return -1;
    }

    int mid = (low + high) / 2;
    if (numbers[mid] == target) {
        return mid;
    } else if (numbers[mid] < target) {
        return binarySearchRecursive(numbers, target, mid + 1, high);
    } else {
        return binarySearchRecursive(numbers, target, low, mid - 1);
    }
}
```

SYNTAX GLOSSARY

Decoding the Java on this page, piece by piece

`if (low > high) { return -1; }` = *the base case, the section to search has shrunk to nothing, so target isn't here*

`return binarySearchRecursive(numbers, target, mid + 1, high);` = *the recursive case, search the smaller right half by calling this exact method again*

`return binarySearchRecursive(numbers, target, low, mid - 1);` = *the recursive case, search the smaller left half instead*

Compare this directly to binarySearchIterative in section 12.3: both compute the same mid, both compare numbers[mid] to target, both narrow the search to one half. The iterative version narrows the search by updating low and high and looping again; the recursive version narrows the search by calling itself with new low and high values instead. Same idea, same base case (low > high, the recursive twin of the loop's low <= high condition), different mechanism.

12.4.1 Call Stack Blueprint: Tracing binarySearchRecursive

Using the same call stack style from Chapter 11, here's binarySearchRecursive searching for the value 7 in the sorted array {1, 3, 5, 7, 9, 11, 13}, positions 0 through 6.

Searching for 7 in {1, 3, 5, 7, 9, 11, 13}

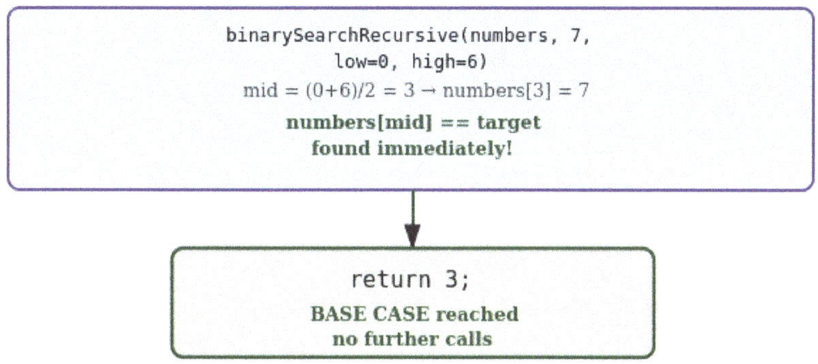

Only one call was ever made. No recursive case was needed this time, because mid landed exactly on the target.

Unlike countDownRecursive in Chapter 11, which always walked all the way down to its base case, this search finds the target on the very first check, mid lands exactly on position 3, where the value 7 lives, and returns immediately, with no further calls at all. This is the entire point of binary search: most searches finish in just a few calls, not many.

12.5 Predict, Observe, Explain (POE): Why Binary Search Needs Order

Before moving to sorting, it's worth testing the one requirement binary search depends on: a sorted array.

PREDICT

Suppose binarySearchIterative is run on an array that is NOT sorted, say, {9, 3, 7, 1, 11, 5, 13}, searching for 7. Do you predict it will still reliably find 7's correct position?

OBSERVE

No. Binary search can fail silently on unsorted data, sometimes reporting "not found" even when the value is present, because eliminating "half" only makes sense if everything on one side is guaranteed smaller and everything on the other side is guaranteed larger. Without that guarantee, low = mid + 1 might eliminate the exact half the target was actually in.

EXPLAIN

In your own words, jot down: why does linearSearch from section 12.2 work correctly on any array, sorted or not, while binarySearch requires sorted data specifically?

This is exactly why sorting, the second half of this chapter, matters so much: it's not just about tidy output. It's what makes fast searching possible in the first place.

12.6 Logical Blueprint: Selection Sort

Here's a plan for sorting a list from smallest to largest, in plain English, before any code: repeatedly find the smallest value remaining in the unsorted part of the list, and swap it into its correct position at the front.

- Look through the entire list and find the position of the smallest value.
- Swap that smallest value into the first position.
- Now treat the rest of the list (everything except the first position) as the new "unsorted part," and repeat, find its smallest value, swap it into the second position.
- Keep repeating, one position at a time, until the unsorted part shrinks to nothing.

Notice the shape: a loop that handles "which position am I filling next," with a smaller search loop inside it to find the right value for that position. This nested-loop pattern is new. The outer loop from Chapter 4 fills one final position per pass, and the inner loop is essentially a linear search for the minimum, from section 12.2.

12.7 Syntax Bridge: Selection Sort

```
public static void selectionSort(int[] numbers) {
    for (int i = 0; i < numbers.length - 1; i = i + 1) {
        int minPosition = i;

        for (int j = i + 1; j < numbers.length; j = j + 1) {
            if (numbers[j] < numbers[minPosition]) {
                minPosition = j;
            }
        }

        int temp = numbers[i];
        numbers[i] = numbers[minPosition];
        numbers[minPosition] = temp;
    }
}
```

🔍 **SYNTAX GLOSSARY**
Decoding the Java on this page, piece by piece

`for (int i = 0; ...) {` = *the outer loop, i is the position currently being filled with its correct value*

`int minPosition = i;` = *assume, for now, that position i already holds the smallest remaining value*

`for (int j = i + 1; ...) {` = *the inner loop, searches only the unsorted part, exactly like section 12.6's plan*

`int temp = ...; numbers[i] = ...; numbers[minPosition] = temp;` = *the standard three-line swap: save one value temporarily so it isn't overwritten before it's used*

Trace through selectionSort on {5, 2, 8, 1} by hand, one outer-loop pass at a time: after the first pass, 1 is swapped to the front, giving {1, 2, 8, 5}. After the second pass, the smallest of the remaining {2, 8, 5}, which is 2, is already in place, so nothing visibly changes. Continue the trace yourself to see the full sort complete.

12.8 Refactor: Choosing the Right Tool

This chapter introduced four algorithms. Refactoring here isn't about rewriting one piece of code. It's about recognizing which tool fits a given situation, the same design judgment from Chapter 7's POE about stored versus computed values.

- Use linearSearch when the data isn't sorted, or you're only searching once. Sorting first just to search once isn't worth the extra work.
- Use binarySearch (either version) when the data is already sorted and you'll search it many times. The speed gain compounds with every additional search.
- Use selectionSort when you need the data sorted at all, regardless of which search you'll use afterward.
- Prefer the iterative version of binary search in most real code, since it avoids the call-stack overhead recursion introduces. But the recursive version is often easier to read and verify as correct, which is exactly why we built both.

There is rarely one "correct" algorithm, only the one that fits the actual situation: how big the data is, how often it changes, and how often you'll search it.

12.9 Practice Problems

Try tracing these by hand first, the way you traced selectionSort in section 12.7, then write or run the Java versions to check yourself.

33. Trace linearSearch and binarySearchIterative searching for 11 in the sorted array {2, 4, 6, 8, 10, 11, 15}. Count how many comparisons each one makes, and explain the difference.

34. Draw a call stack diagram, the way section 12.4.1 did, for binarySearchRecursive searching for 1 in {1, 3, 5, 7, 9, 11, 13}. How many calls does it take this time, and why is it different from the example in 12.4.1?
35. Trace selectionSort on the array {4, 1, 3, 2} completely, writing out the array's contents after every outer-loop pass, the way section 12.7 began.
36. Look back at the POE in section 12.5. A classmate suggests always using binary search because it's faster. Explain, in your own words, why this advice is incomplete, referencing the Refactor section's guidance.

ARCHITECTURAL AI AUDIT

The Scenario: you asked an AI assistant to write a method that checks whether a specific score exists in a list of student scores entered by a teacher throughout the day. It produced the code below. The code is fast and uses the correct algorithm, but it hides a real defect.

```
public static boolean scoreExists(int[] scores, int target) {
    return binarySearchIterative(scores, target) != -1;
}

// Called elsewhere in the program, once per score entered:
int[] todaysScores = {88, 95, 79, 92};
boolean found = scoreExists(todaysScores, 79);
```

TOOLING & WORKFLOW

Augment: For extra practice, generate your own version of this scenario using any AI assistant before comparing it to the one above.

Audit: Do not ask AI to find the defect for you. That defeats the exercise.

Verify: Use the compiler to confirm your fix actually resolves the problem.

13. Find the Defect: Identify the hidden assumption this code depends on. Why does it appear to work in quick testing?
14. Explain the Inconsistency: Describe a realistic, everyday scenario, a teacher entering scores normally, where that hidden assumption breaks silently. Connect your answer to the POE in section 12.5.
15. Correct the Code: Propose a corrected design.
16. Reflect: In one sentence, explain why “this ran correctly in my test” is not the same as “this algorithm fits this situation,” referencing the Refactor section's guidance on choosing tools.

12.10 Chapter Summary

KEY TAKEAWAYS

Linear search checks every element in order and works on any array, sorted or not; it can take as many checks as there are elements.

Binary search eliminates roughly half the remaining data with every check, but requires the array to already be sorted.

Binary search can be written iteratively (a loop narrowing low and high) or recursively (a method calling a smaller version of itself). Both follow the identical logic.

Selection sort repeatedly finds the smallest remaining value and swaps it into place, using a loop inside a loop.

Choosing the right algorithm depends on the situation, data size, how often it's searched, whether it's already sorted, not on always picking the "fastest" one.

12.11 Confidence Checkpoint

Quick gut-check before moving on. These aren't graded, they're just for you, to see what's already sticking.

CHECK YOURSELF

1. In your own words, why does binary search need sorted data, while linear search doesn't?
2. What's the relationship between binarySearchIterative's low <= high and binarySearchRecursive's low > high base case?
3. In selection sort, what does the inner loop actually search for, each time the outer loop runs once?

If any of these feel shaky, that's useful information, not a failure. Flip back to that section before moving to Chapter 13.

Instructor Note: *This chapter intentionally covers only selection sort, not bubble or insertion sort, to keep one clear nested-loop example rather than three similar variations. It also avoids Big-O notation by name, instead building the intuition informally through the phone-book comparison ("roughly 10 checks instead of 1,000"). Formal complexity analysis is left for a follow-up course, since it requires mathematical vocabulary (logarithms) this book has not assumed. Merge sort and quicksort, which themselves use recursion, are natural extensions once this chapter's selection sort and Chapter 11's recursion are both solid, and are suggested as a course capstone or appendix topic.*

CHAPTER 13

Exception Handling

Designing for Failure

THE SIGNPOST: Where We're Headed

In this chapter, we will learn how to anticipate the ways a program can fail at runtime, and respond to those failures on our own terms instead of letting the program crash.

You've already met several of these failures without naming them: an ArrayIndexOutOfBoundsException in Chapter 6, a StackOverflowError in Chapter 11. This chapter gives you a tool to catch failures like these and decide what happens next, instead of the program simply stopping.

13.1 Discovery Challenge: The Crashed Input

Here's a small program, using Scanner from Chapter 5, that asks the user for their age and prints a message back.

```
Scanner keyboard = new Scanner(System.in);
System.out.println("Enter your age:");
int age = keyboard.nextInt();
System.out.println("You are " + age + " years old.");
```

This works perfectly when the user types a number, like 20. But suppose the user types “twenty” instead, a perfectly reasonable, human thing to type, even though it isn't a valid int.

ONE RULE: Read this before you start

You don't need to run this code to answer the question. Reason it through: keyboard.nextInt() expects to read an int. “twenty” is text, not a number Java can convert. Predict, in plain English, what you think happens to the program at the moment the user types “twenty” and presses enter.

TRY IT: Spend about 5 minutes on this

Write one or two sentences predicting what the program's output looks like for a user who types "twenty". Does it print an error message and keep going? Does it print "You are 0 years old"? Does something else happen?

In reality, the program crashes immediately. Java throws an error called InputMismatchException, the line "You are ... years old" never prints at all, and the entire program stops dead, often showing the user a wall of intimidating red text instead of a friendly message.

This is the core discovery of this chapter: plenty of things your program cannot control, what a user types, whether a file exists, whether a network connection succeeds, can cause it to fail partway through, at a moment you can often predict in advance, even if you can't prevent it outright.

13.2 Naming What You Already Did

FORMAL CONCEPT

An unexpected problem that occurs while a program is running is called an exception. When Java encounters one and has no instructions for handling it, the exception is thrown, and the program crashes at that exact point unless something catches it first.

You've actually already met several exceptions by name earlier in this book, without this formal vocabulary: ArrayIndexOutOfBoundsException in Chapter 6, when a loop reached past an array's last valid position, and StackOverflowError in Chapter 11, when a recursive method never reached its base case. InputMismatchException, from this chapter's Discovery Challenge, is the same kind of thing, just triggered by unexpected user input instead of a logic bug.

13.3 Syntax Bridge: try and catch

Here's the age program from section 13.1, rewritten to catch the InputMismatchException instead of letting it crash the program.

```
import java.util.InputMismatchException;
import java.util.Scanner;

Scanner keyboard = new Scanner(System.in);
System.out.println("Enter your age:");

try {
    int age = keyboard.nextInt();
    System.out.println("You are " + age + " years old.");
} catch (InputMismatchException e) {
    System.out.println("That doesn't look like a valid age. Please
enter a number next time.");
}
```

🔍 **SYNTAX GLOSSARY**

Decoding the Java on this page, piece by piece

`try {` = *marks a block of code that might throw an exception, Java watches this block for trouble*

`catch (InputMismatchException e) {` = *if that specific exception happens inside try, run this block instead of crashing*

`e` = *a name for the exception object itself, holding details about what went wrong, in case you need them*

Notice the program no longer crashes for a user who types "twenty." It prints a clear, friendly message and the program continues running normally afterward. The try block is exactly the code that was risky before; the catch block is the plan for what to do if that risk actually happens.

13.3.1 Worked Example: Catching Chapter 6's Array Bug

Recall the ArrayIndexOutOfBoundsException from Chapter 6, caused by reaching past an array's last valid position. The same try/catch pattern handles it too.

```
int[] scores = {88, 92, 79, 95, 84};

try {
    System.out.println(scores[10]);
} catch (ArrayIndexOutOfBoundsException e) {
    System.out.println("That position doesn't exist in this array.");
}
```

Without the try/catch, this would crash exactly the way it did in Chapter 6. With it, the program recognizes the problem, prints something useful, and keeps running. Notice the catch block's parentheses name the specific exception type,

ArrayIndexOutOfBoundsException here, InputMismatchException in section 13.3, each catch block only catches the type it names.

13.3.2 Companion Section: The do-while Loop

The age-reading program from section 13.1 has a real limitation: it catches one bad attempt and prints a friendly message, but then the program simply moves on. It never actually gets a usable age from a user who mistypes. A more realistic program should keep asking until the user provides valid input.

Recall Chapter 4's while loop: it checks its condition before the loop body runs even once. That's normally exactly what you want, but here, there's a chicken-and-egg problem. You can't check whether the input was valid until after you've already asked for it and tried to read it. Java has a loop built for precisely this shape: the do-while loop, which runs its body once first, and only then checks whether to repeat.

```
Scanner keyboard = new Scanner(System.in);
boolean validInput = false;
int age = 0;

do {
    System.out.println("Enter your age:");
    try {
        age = keyboard.nextInt();
        validInput = true;
    } catch (InputMismatchException e) {
        System.out.println("That doesn't look like a number. Try
again.");
        keyboard.nextLine();
    }
} while (!validInput);

System.out.println("You are " + age + " years old.");
```

> 🔍 **SYNTAX GLOSSARY**
> *Decoding the Java on this page, piece by piece*
>
> **`do {`** = *marks the start of a loop body that runs once before any condition is checked*
>
> **`} while (!validInput);`** = *after the body runs, check this condition. If true, run the body again; the loop's check happens at the bottom, not the top*
>
> **`keyboard.nextLine();`** = *clears the leftover bad input left behind after a failed nextInt(), so the next attempt starts clean*

Trace through this by hand for a user who types “twenty,” then “25”: the body runs once, the try fails, validInput stays false, the loop's bottom check sees !validInput is true and repeats. The second time through, "25" succeeds, validInput becomes true, and the loop's bottom check now sees !validInput is false, the loop ends, having run exactly twice.

Compare this to a regular while loop trying to do the same job: a while loop checks validInput before the body ever runs, but validInput doesn't have a sensible starting value until after the user has been asked at least once. do-while solves exactly this ordering problem: run first, then decide whether running again makes sense, and it appears constantly in real programs anywhere user input needs to be validated.

13.3.3 Companion Section: Converting Text to Numbers

Every program in this chapter so far has read a number directly with keyboard.nextInt(), and Java has handled the bad-input checking for us, in the form of InputMismatchException. But not every number starts out as direct keyboard input. Sometimes a program already has a value as a String, maybe read with nextLine(), maybe coming from somewhere else entirely, and needs to convert that text into a usable number afterward.

Java provides a built-in tool for exactly this: Integer.parseInt(), which takes a String and returns the int it represents, if the String actually represents one.

```
String ageText = "25";
int age = Integer.parseInt(ageText);
System.out.println(age + 1);
```

🔍 **SYNTAX GLOSSARY**
Decoding the Java on this page, piece by piece

`Integer.parseInt(ageText)` = *convert the String ageText into an int, if its text actually represents one*

This works perfectly for "25". But suppose ageText held "twenty-five" instead, the exact same bad input from section 13.1, just arriving as a String instead of straight from nextInt(). Integer.parseInt() cannot make sense of that text, and it fails, the same way nextInt() failed before.

The exception it throws has a different name, though: not InputMismatchException, but NumberFormatException. Different tool, different exception type, same underlying idea: Java refuses to silently guess at a bad conversion, and tells you immediately instead.

```
try {
    int age = Integer.parseInt(ageText);
    System.out.println("You are " + age + " years old.");
} catch (NumberFormatException e) {
    System.out.println("That doesn't look like a number.");
}
```

Notice the shape of this try/catch is identical to every other one in this chapter, only the exception type named in catch changed. This is the real, reusable pattern of this entire chapter: identify what could go wrong, name the specific exception Java throws for it, and respond instead of crashing.

13.4 Practice Problems

Try predicting the behavior by hand first, the way you predicted the crashed input in section 13.1, then write or run the Java versions to check yourself.

- Wrap Chapter 8's setGpa() validation logic in your head: if a setter already rejects invalid values with an if/else (no exception involved), why might try/catch still not be the right tool there? Connect your answer to the difference between a value that's simply out of range and a genuinely unexpected failure like bad user input.
- Write a try/catch around a line that divides by a variable which might be zero, catching ArithmeticException, the exception Java throws for integer division by zero. What message would you print in the catch block?
- A program reads a user's age with keyboard.nextInt() inside a try block, catching InputMismatchException. Trace through what happens for three different users typing: "25", "twenty-five", and "-5". Which of these actually triggers the catch block, and which doesn't, even though it's also a strange input?
- Look back at the Discovery Challenge in section 13.1. Explain, in your own words, why catching an exception is different from preventing it. The program still encountered the bad input, it just didn't crash because of it.
- Trace through the do-while program in section 13.3.2 for a user who types "abc", then "xyz", then "30". How many times does the loop body run, and what is the final value of age?
- Write a try/catch around Integer.parseInt() that converts a String variable called priceText into an int, catching NumberFormatException. What would you print in the catch block, and how is this similar to catching InputMismatchException in Practice Problem 3?

Optional: if you used an AI assistant to write a try/catch here, check that it named the correct exception type yourself, the way this chapter paired each tool with its specific failure. AI-generated code sometimes catches a plausible-sounding exception that isn't actually the one the code can throw.

13.5 Chapter Summary

KEY TAKEAWAYS

An exception is an unexpected problem that occurs while a program is running; an uncaught exception crashes the program at that exact point.

try marks code that might fail; catch names a specific exception type and provides a plan for what to do if it happens.

ArrayIndexOutOfBoundsException (Chapter 6) and StackOverflowError (Chapter 11) are both exceptions you've already met. try/catch can handle the first, though a stack overflow is usually better prevented with a correct base case.

Catching an exception doesn't prevent the underlying problem. It lets your program respond to it gracefully instead of crashing.

13.6 Confidence Checkpoint

Quick gut-check before moving on. These aren't graded, they're just for you, to see what's already sticking.

CHECK YOURSELF

1. In your own words, what is an exception?
2. What's the difference between the try block and the catch block?
3. Name one exception you met earlier in this book, before this chapter gave it a formal name.

If any of these feel shaky, that's useful information, not a failure. Flip back to that section before moving to Chapter 14.

Instructor Note: *Section 13.3.2 resolves a promise made in Chapter 4's instructor note: do-while was deliberately withheld until a genuine motivating use case existed, and input validation, try once, check the result, repeat if needed, is exactly that case. Section 13.3.3 was added for a related reason: Integer.parseInt() and NumberFormatException are common enough in real Java code that leaving them entirely to an appendix risked under-serving students who go on to a second course, while the do-while companion section already established the right scope and tone for a short, motivated addition late in the chapter. This chapter otherwise intentionally keeps Light Arc scope: try/catch with a single catch block, on exceptions students have already encountered informally. finally blocks, multi-catch syntax, custom exception classes, and the checked-versus-unchecked distinction are all deliberately deferred. They add real value for larger programs but aren't needed to build the core mental model of "anticipate failure, respond instead of crash." Practice Problem 1 is worth discussing directly: validation with if/else (Chapter 8's approach) and try/catch (this chapter's approach) solve related but distinct problems. The former prevents bad state from being set in the first place, the latter responds to failures that can't be prevented in advance, like malformed user input.*

CHAPTER 14

Object-Oriented System Design (OOP)

Bringing everything together in one working system.

THE SIGNPOST: Where We're Headed

In this chapter, we will bring together everything from this book, decomposition, state, decisions, loops, objects, arrays, classes, encapsulation, inheritance, polymorphism, recursion, and exception handling, to build one complete, working system.

This chapter doesn't introduce new syntax. It's a capstone: a chance to see how every tool from Chapters 1 through 13 fits together in a single real program, the way they actually do in professional software, not as thirteen separate ideas, but as one coherent system.

14.1 Discovery Challenge: Designing the Roster System

Throughout this book, you've built pieces of a student records system without ever assembling them: a Student class (Chapter 7), encapsulated GPA validation (Chapter 8), a GraduateStudent that extends it (Chapter 9), polymorphic reporting (Chapter 10), and exception-safe input (Chapter 13).

Before writing any new code, sketch on paper what a complete Roster Management System needs, pulling only from tools you already have. What class holds the collection of students? What operations should it support: adding a student, searching for one by ID, printing a full report?

ONE RULE: Read this before you start

Every piece of your design must trace back to a specific earlier chapter. No new concepts allowed. If you find yourself wanting something this book hasn't covered, simplify your design until it only uses tools you already have. This is what real system design feels like: working within real constraints, not waiting for a perfect, unlimited toolkit.

TRY IT: Spend about 10 minutes on this

Sketch a CRC card (Chapter 7's tool) for a new class, Roster, that holds many Student objects. List its fields and methods, and next to each one, write which earlier chapter's concept it relies on.

A reasonable Roster design holds an array of Student objects (Chapter 6) as a private field (Chapter 8), with methods to add a student, search for one by ID (Chapter 12's linear search, since student records aren't naturally sorted by ID order), and print a full report by looping through and calling reportGpa() on each one, which, thanks to polymorphism (Chapter 10), correctly handles both Student and GraduateStudent objects without any special-casing.

This is the core discovery of this chapter: a real system isn't one new big idea. It's a careful assembly of small, well-understood pieces, each one doing exactly the job it was designed for back when you first learned it.

14.2 Logical Blueprint: The Roster Class

Here's a CRC card for Roster, matching the discovery above, mapped explicitly to the chapters each piece depends on:

Class: Roster

Fields:

- students: a private array of Student objects (Chapters 6, 8)
- count: a private int tracking how many students are currently stored (Chapter 2)

Methods:

- addStudent(Student s): stores a new student in the next open array position (Chapters 2, 6)
- findById(int id): linear search through students for a matching studentId, returning null if not found (Chapter 12)
- printReport(): loops through students and calls reportGpa() on each, relying on polymorphism to handle Student and GraduateStudent correctly (Chapters 4, 10)

Notice this card is, in spirit, no different from the Student card you built in Chapter 7, information it holds, actions it performs. System design at this scale is the same skill you've already practiced, just applied to a class that manages other objects instead of simple values.

Here's that same design as a diagram, showing how Roster, Student, and GraduateStudent connect, with each piece tagged by the chapter where you first built it.

The Roster System — Where Each Piece Came From

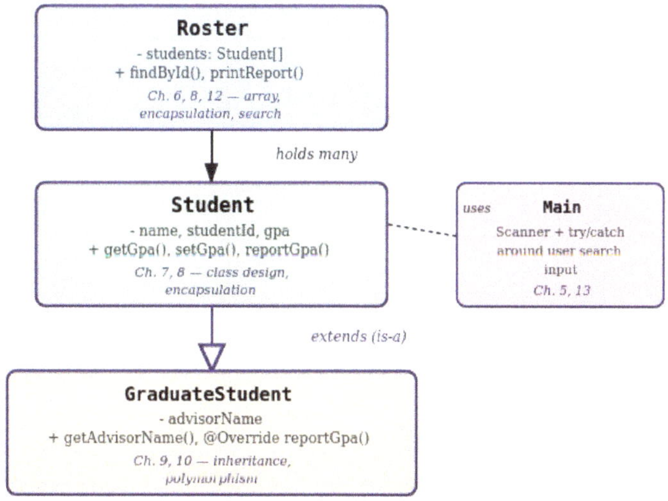

The solid arrow means "holds a collection of." The hollow arrow means "is a more specific version of." Every box and connection in this diagram is a tool you already built, in an earlier chapter.

Notice the two different arrow styles: a solid arrow means "this class holds a collection of these," the relationship between Roster and Student. A hollow-triangle arrow means "is a more specific version of," the exact is-a relationship from Chapter 9. GraduateStudent extends Student. These two arrow styles cover most of the structural relationships you'll encounter in real object-oriented systems.

14.3 Syntax Bridge: Assembling the Roster Class

Here's the Roster class, built from the Blueprint above, using the Student and GraduateStudent classes exactly as they were left at the end of Chapter 10.

```
public class Roster {
    private Student[] students;
    private int count;

    Roster(int capacity) {
        students = new Student[capacity];
        count = 0;
    }

    public void addStudent(Student s) {
        if (count < students.length) {
            students[count] = s;
            count = count + 1;
        } else {
            System.out.println("Roster is full, cannot add more
students.");
        }
    }

    public Student findById(int id) {
        for (int i = 0; i < count; i = i + 1) {
            if (students[i].getStudentId() == id) {
                return students[i];
            }
        }
        return null;
    }

    public void printReport() {
        for (int i = 0; i < count; i = i + 1) {
            students[i].reportGpa();
        }
    }
}
```

🔍 **SYNTAX GLOSSARY**
Decoding the Java on this page, piece by piece

`private Student[] students;` = *Chapter 6's array, holding Chapter 9's class hierarchy, kept private per Chapter 8*

`for (int i = 0; i < count; ...) {` = *loops only through filled positions, not the array's full capacity, count tracks how many are actually in use*

`students[i].reportGpa();` = *Chapter 10's polymorphism, this single line correctly handles every kind of Student stored here*

`return null;` = *a special value meaning "no object here," the method's way of saying findById found nothing*

Notice findById uses Chapter 12's linear search pattern exactly, and printReport uses Chapter 4's loop combined with Chapter 10's polymorphism. One line, students[i].reportGpa(), correctly produces different output for a Student versus a GraduateStudent, with zero type-checking code anywhere in this class.

14.3.1 Worked Example: A Complete Program

Here's a full Main class exercising the Roster, including exception-safe input from Chapter 13 for a search feature.

```java
import java.util.InputMismatchException;
import java.util.Scanner;

public class Main {
    public static void main(String[] args) {
        Roster roster = new Roster(10);
        roster.addStudent(new Student("Maria", 1002, 3.6));
        roster.addStudent(new GraduateStudent("Ada", 1001, 3.9, "Dr.
Turing"));

        roster.printReport();

        Scanner keyboard = new Scanner(System.in);
        System.out.println("Enter a student ID to search:");

        try {
            int searchId = keyboard.nextInt();
            Student found = roster.findById(searchId);
            if (found != null) {
                found.reportGpa();
            } else {
                System.out.println("No student found with that ID.");
            }
        } catch (InputMismatchException e) {
            System.out.println("Please enter a valid numeric ID.");
        }
    }
}
```

Every line of this program traces back to a specific earlier chapter: object creation (5, 7), inheritance (9), arrays wrapped in a class (6, 8), polymorphism (10), conditional logic checking for null (3), and exception handling around user input (13). Nothing here is new. This chapter's entire job was showing you that you already had everything needed to build this.

14.4 Practice Problems

These are larger than previous chapters' practice problems on purpose. Treat them as a capstone project, not a quick exercise.

37. Add a removeById(int id) method to Roster that removes a student by shifting every later element one position to the left, then decrementing count. Trace through your method by hand on a 3-student roster before writing the Java.
38. Add a sortByGpa() method to Roster using Chapter 12's selectionSort pattern, adapted to compare students[i].getGpa() instead of plain int values.

39. Extend the ExchangeStudent class from Chapter 9's practice problems so it can be added to a Roster alongside Student and GraduateStudent objects. What, if anything, needs to change in the Roster class itself to support this third type? Connect your answer to Chapter 10's POE about scaling type-checking code.
40. Design (CRC-card style, the way section 14.2 did) a second manager class, CourseRoster, that holds a collection of a new Course class. Identify at least two design decisions you'd make the same way as Roster, and one you'd make differently, with justification.

Optional: if you used an AI assistant on any of these capstone problems, run the same kind of audit on its output that you've now practiced five times in this book, find the design principle at stake, trace the logic by hand, and verify with the compiler before trusting it.

14.4.1 Companion Section: ArrayList

Practice Problem 1 just asked you to write a removeById method that shifts every element after the removed one a position to the left and decrements count. That shifting logic is real, correct, and exactly how a fixed-size array manager has to work. It is also genuinely tedious to write, easy to get wrong, and something Java's standard library already handles for you.

ArrayList is a built-in Java class that manages a resizable list of objects. It does the shifting, the resizing, and the bookkeeping automatically. A Roster built on ArrayList would not need a fixed capacity, a separate count variable, or any manual shifting logic. The list grows as you add objects and shrinks as you remove them.

```
import java.util.ArrayList;

ArrayList<Student> students = new ArrayList<Student>();
```

🔍 **SYNTAX GLOSSARY**
Decoding the Java on this page, piece by piece

`import java.util.ArrayList;` = *required at the top of any file that uses ArrayList, same pattern as Scanner*

`ArrayList<Student>` = *an ArrayList that holds Student objects, the angle brackets name the type it stores*

`new ArrayList<Student>()` = *creates an empty list, ready to grow as you add objects*

Once created, an ArrayList gives you five methods the AP Java Quick Reference guarantees will be on the exam. Each one replaces something you have been doing manually in the Roster class.

```
students.add(new Student("Ada", 1001, 3.9));
int count = students.size();
Student s = students.get(0);
students.set(0, new Student("Alan", 1002, 3.7));
students.remove(0);
```

🔍 SYNTAX GLOSSARY

Decoding the Java on this page, piece by piece

`students.add(object)` = *appends an object to the end of the list, list grows automatically*

`students.size()` = *returns the number of objects currently in the list, same role as count in Roster*

`students.get(i)` = *returns the object at position i, same as students[i] in the array version*

`students.set(i, object)` = *replaces the object at position i with a new one*

`students.remove(i)` = *removes the object at position i and shifts everything after it left automatically*

Looping through an ArrayList uses the same for loop pattern as an array, with size() replacing length.

```
for (int i = 0; i < students.size(); i = i + 1) {
    students.get(i).reportGpa();
}
```

Java also provides a shorter loop syntax called the for-each loop that works on both arrays and ArrayLists. Instead of managing an index variable yourself, you name a variable that takes on each value in the list, one at a time.

```
for (Student s : students) {
    s.reportGpa();
}
```

🔍 SYNTAX GLOSSARY

Decoding the Java on this page, piece by piece

`for (Student s : students)` = *for each Student in students, call it s, runs the body once per object in the list*

ONE RULE: before you move on

ArrayList stores objects, not primitives. ArrayList<int> is not valid Java. To store numbers in an ArrayList, use the wrapper class: ArrayList<Integer> for whole numbers, ArrayList<Double> for decimals. This is why Integer and Double exist as class names alongside the primitive int and double.

To see the difference concretely, here is the addStudent logic from section 14.3, first with the fixed array, then with ArrayList.

```
// Fixed array version (from section 14.3)
public void addStudent(Student s) {
    if (count < students.length) {
        students[count] = s;
        count = count + 1;
    }
}

// ArrayList version
public void addStudent(Student s) {
    students.add(s);
}
```

The ArrayList version is not smarter, it delegates the bookkeeping to a class that was already written and tested. This is the same idea behind using Scanner instead of writing your own keyboard-reading code in Chapter 5: Java's standard library exists so you do not have to rebuild solved problems from scratch.

41. Rewrite the Roster class from section 14.3 using ArrayList<Student> instead of a fixed array. Remove the capacity constant and the count variable. Which methods become shorter? Which stay the same length?
42. Write a for-each loop that prints the name and GPA of every student in an ArrayList<Student> called roster. Then rewrite the same loop using a standard index-based for loop. Which is clearer for this task?

Optional: if you used an AI assistant here, check that its ArrayList methods match the five on the AP Java Quick Reference exactly. AI-generated ArrayList code often uses methods that exist in the full Java library but are not on the Quick Reference and therefore cannot be assumed on the AP exam.

KEY TAKEAWAYS

System design is the assembly of small, well-understood pieces, not a single new big idea.

A manager class (like Roster) commonly combines a private array (Chapter 6, 8) with methods built from earlier patterns: loops (Chapter 4), search (Chapter 12), and polymorphic calls (Chapter 10).

Designing a class that manages other objects uses the exact same CRC-card process as designing any other class, from Chapter 7.

A complete, real program touches nearly every concept in this book at once. The skill of this course was never any one chapter alone, but learning to combine them.

14.6 Confidence Checkpoint

This is the last checkpoint in the book. Rather than three narrow questions, take a moment with one bigger one.

CHECK YOURSELF

1. Pick any one method from the Roster class in section 14.3. Trace its logic back to the specific chapter where you first learned that pattern, and explain the connection in your own words.

If you can do that for any method in this chapter, you've learned what this book set out to teach: not isolated facts about Java, but a way of thinking that breaks problems down, builds them back up, and recognizes the same patterns wherever they reappear.

14.7 A Closing Note: The Skill You've Been Building

There is a thread that runs through this entire book that is worth naming one last time before you put it down.

Five times across these fourteen chapters, you encountered an Architectural AI Audit. Each one placed you in the same position: an AI had produced code that compiled, ran, and looked reasonable. Your job was not to accept it. Your job was to find the flaw, understand why it mattered, and build something better.

In Chapter 6, you found an off-by-one boundary error that only surfaces on the last element. In Chapter 8, you found a field left unprotected in an otherwise encapsulated class. In Chapter 9, you identified an inheritance relationship that the is-a test immediately disqualified. In Chapter 10, you traced an instanceof check that the language's own polymorphism had already made unnecessary. In Chapter 12, you caught a hidden assumption about sorted input that would silently corrupt results in normal use.

Beyond those five chapters, every practice problem section in this book ended with the same optional reminder, worded differently each time, but always pointing at the same habit: use the verification skill this chapter just taught you, not your confidence in an AI's answer, as the actual test of whether a solution is correct.

THE SIGNPOST: where you have arrived

You have now practiced this five times as a named, structured exercise and reinforced it across nine more chapters as an everyday habit. In professional software development, this combination has a name. Using AI as a productive partner in the act of creation is called AI-Augmented practice. Maintaining your own judgment as the final authority on correctness and design is called AI-Resilience. Both are skills the industry now specifically looks for in new developers. You did not learn them at the end of this book. You built them across the whole thing.

The Architect does not stop being the Architect because the Draftsman got faster.

Instructor Note: *This capstone is designed to be assigned as a multi-day or multi-week project rather than a single sitting. The Practice Problems are intentionally open-ended and larger in scope than any prior chapter, suitable as a final assessment. Instructors may wish to let students choose their own domain (a library system, a game inventory, a recipe box) instead of the Roster example, as long as their design follows the same Blueprint process in section 14.2, a manager class wrapping a private collection, built from patterns established in Chapters 1-13. This is also a natural point to introduce ArrayList, ArrayList<Student> in particular, as a follow-up enhancement once students see firsthand the friction of Roster's fixed-capacity array (Practice Problem 1's shifting-elements logic is exactly the kind of bookkeeping ArrayList exists to eliminate).*

APPENDIX A

Java Quick Reference

Every syntax pattern from this book, in one place

HOW TO USE THIS APPENDIX

This is a lookup tool, not a teaching tool. Every pattern here was taught in full, with a worked example, somewhere in Chapters 1 through 14. Use this appendix to quickly recall exact syntax while you're working on an assignment. If a pattern doesn't make sense, the chapter reference will take you back to where it was actually explained.

The first two tables group related syntax by category, useful when you remember what kind of thing you're looking for (a data type, a loop) but not which chapter taught it. Every table after that is grouped by chapter, in the order you learned it.

A.0 Data Types

Syntax	Explanation	Ch.
`int`	a whole number, no decimal point	2
`double`	a decimal number	2
`String`	text, held in an object with its own methods	2
`boolean`	only ever true or false	3

A.0 Control Flow

Syntax	Explanation	Ch.
`if (...) { } else { }`	branch on a condition, two paths	3
`else if (...) { }`	check a second condition if the first was false	3
`switch (...) { case: }`	branch on one variable against exact values	7
`while (...) { }`	loop, condition checked before each pass	4
`for (...; ...; ...) { }`	loop, start/check/update gathered into one line	4

Syntax	Explanation	Ch.
`do { } while (...);`	loop, runs once before the condition is ever checked	13
`break;`	exit a loop or switch case early	7
`continue;`	skip the rest of this pass, move to the next one	7

A.1 Variables and Output (Chapter 2)

Syntax	Explanation	Ch.
`int cashBox = 0;`	declare a whole-number variable, starting at 0	2
`double gpa = 3.9;`	declare a decimal-number variable	2
`String name = "Ada";`	declare a text variable	2
`boolean onTheList = true;`	declare a true/false variable	2
`cashBox = cashBox + 5;`	update a variable: read it, change it, store it back	2
`System.out.println(cashBox);`	print a value, followed by a new line	2
`// a comment`	plain-English notes for the reader, ignored by Java	2

A.2 Selection Structures (Chapter 3)

Syntax	Explanation	Ch.
`if (onTheList) { }`	run the block only if the condition is true	3
`} else { }`	run this block instead, if the condition was false	3
`} else if (hasGuardian) { }`	check a second condition, only if the first was false	3
`age >= 18`	comparison operators: >=, <=, >, <, ==, !=	3

A.3 Looping Structures (Chapter 4)

Syntax	Explanation	Ch.
`while (count < 100) { }`	a while loop: check first, run if true	4
`for (int i = 0; i < 100; i++) { }`	same loop, all three parts in one line	4

A.4 Objects, Scanner, and String (Chapter 5)

Syntax	Explanation	Ch.
`import java.util.Scanner;`	required at the top of a file before Scanner can be used	5
`Scanner keyboard = new Scanner(System.in);`	create a Scanner that reads keyboard input	5
`keyboard.nextLine()`	read one full line the user types, as a String	5
`keyboard.nextInt()`	read one whole number the user types	5
`name.length()`	ask a String how many characters it holds	5
`name.toUpperCase()`	ask a String for an all-uppercase version	5
`name.indexOf("text")`	ask a String where "text" first appears	5
`name.substring(1, 3)`	return characters from position 1 up to but not including position 3	5
`name.charAt(0)`	return the single character at position 0	5
`name.equals(other)`	return true if both Strings hold the same text (use this, not ==)	5

A.5 Arrays (Chapter 6)

Syntax	Explanation	Ch.
`int[] scores = {88, 92, 79};`	declare an array with starting values	6
`scores.length`	the number of positions in the array	6
`scores[2]`	read the value at position 2 (positions start at 0)	6
`scores[2] = 84;`	update the value at position 2	6
`for (int i = 0; i < scores.length; i++) { }`	the standard, safe pattern for looping through an array	6
`int[][] grid = {{1,2},{3,4}};`	declare a 2D array (rows and columns)	6
`grid[1][2]`	read row 1, column 2 (row first, column second)	6
`grid.length`	the number of rows in a 2D array	6
`grid[0].length`	the number of columns in a 2D array	6
`for (int r=0; r<grid.length; r++) { for (int c=0; c<grid[r].length; c++) { } }`	nested loop traversal of a 2D array	6

A.6 Class Design (Chapter 7)

Syntax	Explanation	Ch.
`public class Student { }`	begin a class definition	7
`String name; int studentId;`	fields: the information a class holds	7
`Student(String n, int id) { }`	a constructor, runs once when an object is created	7
`void introduce() { }`	a method with no return value	7
`Student ada = new Student("Ada", 1001);`	create an object from a class	7
`ada.introduce();`	call a method on an object	7

A.7 Encapsulation (Chapter 8)

Syntax	Explanation	Ch.
`private double gpa;`	a field unreachable directly from outside the class	8
`public double getGpa() { }`	a getter: the safe, public way to read a private field	8
`public void setGpa(double g) { }`	a setter: the safe, public way to update a field, with validation	8
`newGpa >= 0.0 && newGpa <= 4.0`	&&: both conditions must be true	8

A.8 Inheritance (Chapter 9)

Syntax	Explanation	Ch.
`class GraduateStudent extends Student { }`	declare an is-a relationship	9
`super(name, id, gpa);`	call the parent class's constructor first	9
`this.advisorName = advisor;`	this refers to the object being built right now	9

A.9 Polymorphism (Chapter 10)

Syntax	Explanation	Ch.
`@Override`	marks a method as intentionally replacing an inherited one	10
`void reportGpa() { }`	redefining a method with the same name and parameters as the parent's version	10

A.10 Recursion (Chapter 11)

Syntax	Explanation	Ch.
`if (n > 0) { }`	the recursive case check: keep going while there's work left	11
`countDownRecursive(n - 1);`	a method calling itself, with a smaller input each time	11

Every recursive method needs a base case (where it stops) and a recursive case (where it makes progress toward that base case).

A.11 Search and Sort (Chapter 12)

Syntax	Explanation	Ch.
`return -1;`	a signal meaning "not found"	12
`int low = 0; int high = arr.length - 1;`	binary search boundaries: the section still worth checking	12
`int mid = (low + high) / 2;`	the middle position of the current section	12
`if (low > high) { return -1; }`	binary search's base case: nothing left to check	12

A.12 Exception Handling (Chapter 13)

Syntax	Explanation	Ch.
`try { }`	marks a block of code that might throw an exception	13
`catch (InputMismatchException e) { }`	catches one specific exception type, runs instead of crashing	13
`do { } while (!validInput);`	runs its body once first, then checks whether to repeat	13
`Integer.parseInt(ageText)`	convert a String into an int, if its text actually represents one	13
`catch (NumberFormatException e) { }`	catches a failed Integer.parseInt() conversion	13

A.13 Putting It Together (Chapter 14)

Syntax	Explanation	Ch.
`private Student[] students;`	an array wrapped inside a class	14
`for (int i = 0; i < count; i++) { }`	looping only through filled positions	14

Syntax	Explanation	Ch.
`students[i].reportGpa();`	calling a method polymorphically inside a loop	14
`import java.util.ArrayList;`	required before ArrayList can be used	14
`ArrayList<Student> list = new ArrayList<Student>();`	create a resizable list of Student objects	14
`list.add(s);`	append an object to the end of the list	14
`list.size()`	the number of objects currently in the list	14
`list.get(i)`	return the object at position i	14
`list.set(i, s);`	replace the object at position i	14
`list.remove(i);`	remove the object at position i, shifts remainder left	14
`for (Student s : list) { }`	for-each loop: runs once per object in the list	14

APPENDIX B

Common Runtime Error Guide

What to do when your program crashes

HOW TO USE THIS APPENDIX

When a Java program crashes, it prints a wall of red text called a stack trace. Buried in that text is the actual error type, a name ending in Exception or Error. Find that name in this appendix to understand what went wrong and how to fix it.

This appendix has two parts. Section B.1 covers the four errors this book actually taught, each tagged with the chapter where you can find a full worked example. Section B.2 is a bonus reference: errors you're likely to encounter in future coursework or personal projects that this book did not teach. Section B.2 is genuinely new material, not a quiz on anything covered here. Read it only if you're curious or you've already hit one of these errors and landed on this page looking for help.

B.1 Errors Covered in This Book

Each of these was taught with a full worked example. If the short explanation here doesn't fully click, the chapter tag tells you exactly where to go for the complete walkthrough.

`ArrayIndexOutOfBoundsException` *(Chapter 6, section 6.4)*

What it means: Your code tried to access an array position that doesn't exist.

Common cause: Using <= instead of < in a loop condition (for example, i <= scores.length instead of i < scores.length), or simply guessing a position number that's outside the array's valid range.

Fix: Remember that array positions start at 0, and the last valid position is always arrayName.length - 1. Double-check any loop condition that touches an array.

`StackOverflowError` *(Chapter 11, section 11.5)*

What it means: A recursive method called itself too many times without stopping, and Java ran out of space to keep track of all the waiting calls.

Common cause: A missing base case, or a base case that's never actually reached because the recursive case doesn't make genuine progress toward it.

Fix: Check that your recursive method has a clear base case, and that every recursive call moves strictly closer to that base case (for example, n - 1, not n or n + 1).

`InputMismatchException` *(Chapter 13, sections 13.3 and 13.3.2)*

What it means: Your program asked for one kind of input (like a number) using Scanner, but the user typed something that doesn't match.

Common cause: Calling keyboard.nextInt() when the user types letters or symbols instead of digits.

Fix: Wrap the risky read in a try/catch block. If you need to keep asking until the input is valid, pair the try/catch with a do-while loop.

`ArithmeticException` *(Chapter 13, Practice Problem 2)*

What it means: Your code tried to perform an operation that's mathematically undefined, most commonly dividing an integer by zero.

Common cause: A divisor variable that turned out to be 0 at the moment the division actually ran, often because it depended on user input or an earlier calculation that wasn't validated.

Fix: Check whether a divisor could be zero before dividing, or wrap the division in a try/catch that catches ArithmeticException.

B.2 Beyond This Book

A NOTE BEFORE YOU READ THIS SECTION

Everything below this point goes beyond what this book taught. These are common errors in real Java programs that use tools outside this book's scope (ArrayList, type casting, file I/O). Treat this section as a head start for whatever you study next, not a review of material you're expected to already know.

`NullPointerException`

What it means: Your code tried to use an object that doesn't actually exist yet. It points to nothing (null) instead of a real object.

Common cause: Calling a method on a variable that was declared but never assigned an actual object, or that was the result of a search method (like this book's findById) that returned null because nothing matched.

Fix: Before calling a method on an object that might not exist, check it isn't null first, the same null-check pattern used in Chapter 14's Main class.

ClassCastException

What it means: Your code tried to treat an object as a more specific type than it actually is.

Common cause: Forcing a conversion between unrelated types, most often when working with collections that hold a general type but are assumed to secretly hold something more specific.

Fix: Use instanceof to check an object's actual type before casting it, or redesign so the cast isn't needed at all.

NumberFormatException

What it means: Your code tried to convert text into a number, but the text wasn't a valid number.

Common cause: Calling Integer.parseInt() or Double.parseDouble() on a String that contains letters, symbols, or is empty.

Fix: Validate the String's content before converting it, or wrap the conversion in a try/catch.

ConcurrentModificationException

What it means: Your code tried to add or remove items from a collection while a loop was actively iterating over that same collection.

Common cause: Removing an item from inside a for-each loop over an ArrayList, instead of using a standard indexed loop or an Iterator built for safe removal.

Fix: Avoid modifying a collection's size while looping over it directly. Build a separate list of changes and apply them after the loop finishes.

FileNotFoundException

What it means: Your code tried to open a file that doesn't exist at the path it was given.

Common cause: A typo in the file path, or a path that's relative to the wrong starting folder.

Fix: Print the exact path your code is trying to open and verify it manually. Use an absolute path while debugging to rule out folder confusion.

APPENDIX C

AI-Augmented, AI-Resilient Pedagogy

For instructors, researchers, and curriculum reviewers

C.1 Purpose of This Appendix

This appendix is addressed to instructors adopting this text, researchers studying AI integration in computing education, and curriculum reviewers evaluating the book's pedagogical claims. It is not student-facing material. Its purpose is to make the book's AI-related pedagogical framework explicit, locatable, and citable, rather than leaving it to be inferred from individual exercises scattered across fourteen chapters.

The framework has a name: the AI-Augmented, AI-Resilient Professional Loop. The following sections define each component, explain its positioning within the existing research literature, and identify exactly where each component recurs in the text.

C.2 Framework Definitions

AI-Augmented Pedagogy

Deliberately weaving AI into tasks to enhance human capabilities and share the cognitive load, while maintaining the student as the director of intent. In this framework, AI functions as the Draftsman: a capable producer of first-pass solutions that the student, acting as Architect, then evaluates and improves. AI-Augmented practice does not discourage AI use; it structures it.

This usage aligns with the term "AI-augmented workflows" as used in industry contexts and with the broader "AI-augmented learning" framing now established in higher education literature. Supporting data: 85% of professional developers regularly use AI coding tools, with 51% using them daily (JetBrains, 2025; Stack Overflow, 2025; Modall, 2025).

AI-Resilient Pedagogy

Designing assessments and exercises that protect critical thinking, ensuring students cannot easily bypass learning via AI shortcuts. An AI-resilient exercise is not one that is impossible to complete with AI assistance. It is one where AI assistance cannot substitute for genuine understanding, because the skill being assessed requires reasoning the student must perform themselves.

This usage matches the term as it appears in peer-reviewed computing education research. Ding (2024) presents an empirically validated framework for "AI-resilient assessment design" in computing and data science courses, documenting a 30-percentage-point drop between AI-assisted homework performance and proctored exam performance (Cohen's d = 1.51) as evidence of the failure mode this framework addresses. Akbar (2025) frames the same concern at the University of Sydney: detection-based approaches are fundamentally inadequate, and the only durable solution is assessment design that makes AI substitution structurally impossible.

C.3 The Professional Loop: A Three-Role Framework

This textbook operationalizes both concepts through a three-role sequence repeated five times across the book as a named exercise (the Architectural AI Audit) and reinforced in abbreviated form across nine additional chapters.

Role	Function	Framework Component
The Draftsman (AI)	Generates syntactically correct, often convincing first-pass code. In several Audit exercises, this code compiles and runs without error; this is intentional.	*AI-Augmented*
The Architect (Student)	Reviews the Draftsman's output, identifies the design or correctness flaw, and reasons out an improved structure, without delegating this evaluation back to AI.	*AI-Resilient*
The Workshop (Compiler)	Confirms that the student's corrected implementation is syntactically valid and behaviorally correct. Notably, the compiler verifies correctness, not architectural soundness; that judgment remains the Architect's.	*Verification (neutral)*

C.4 Where the Framework Recurs in This Text

Full Architectural AI Audits (5 chapters)

Each Audit presents a complete Professional Loop exercise. Two variants were developed to match the actual nature of each chapter's flaw:

- Design-Flaw variant (Chapters 9, 10): used when the AI-generated code is syntactically and behaviorally correct but represents a genuine architectural liability (wrong is-a relationship; unnecessary instanceof defeating polymorphism). Students apply Debt Analysis, Fragility Risk, Refactor, and Reflect tasks.

- Bug-Audit variant (Chapters 6, 8, 12): used when the AI-generated code contains a correctness error or a hidden assumption. Students apply Find the Defect, Explain the Inconsistency, Correct the Code, and Reflect tasks.

AI-Resilience Reminders (8 additional chapters)

Chapters 2, 3, 4, 5, 7, 11, 13, and 14 each close their practice problem section with a one-line, chapter-specific reminder. Each reminder names the verification skill the chapter just taught (state tracing, branch tracing, boundary checking, base case identification, and so on) and explicitly frames AI's confidence as separate from correctness. Chapter 14's reminder invites students to apply the full Audit pattern to their own capstone work, unprompted.

Tooling and Workflow Callout (all 5 Audit chapters)

A short gold callout box at the top of each Audit section names the three phases explicitly: Augment (use AI to generate the scenario), Audit (do not ask AI to evaluate it, that defeats the exercise), and Verify (use the compiler to confirm your fix). This callout uses chapter-specific wording for the Verify instruction to match the nature of each Audit's flaw.

C.5 The Dual Demand: Why Both Terms Are Necessary

The terms AI-Resilient and AI-Augmented are not interchangeable, and their pairing is not rhetorical. They name two distinct failure modes this framework is designed to prevent simultaneously.

- AI over-reliance (the AI-Resilient concern): students who accept AI-generated code uncritically, treating confidence as a proxy for correctness. This is the failure mode documented by Ding (2024) and Akbar (2025). The five Architectural AI Audits directly address this by making the detection of plausible-but-flawed AI output the assessed skill itself.
- AI avoidance (the AI-Augmented concern): students who refuse to engage with a tool that is already standard practice in the software industry. As of 2025, 85% of professional developers regularly use AI tools for coding (JetBrains, 2025), and developers who can effectively direct and review AI-generated output are becoming more valuable than those who write every line manually (Modall, 2025). A curriculum that does not prepare students for this reality is leaving them unprepared for the workforce they are entering.

A curriculum that addresses only one of these fails on the other axis. The Professional Loop addresses both in the same exercise: the Draftsman step normalizes AI use; the Architect step requires independent judgment; the Workshop step provides an objective verifier that is neither the student nor the AI.

C.6 Citable Summary Statement

This textbook implements a dual-strand AI pedagogy framework across all fourteen chapters: an AI-Augmented strand (the Professional Loop's Draftsman and Architect roles, instantiated as five full Architectural AI Audit exercises) and an AI-Resilient strand (chapter-specific verification reminders in eight additional practice problem sections). Together, these produce a student who is neither discouraged from using AI as a professional tool nor permitted to substitute AI output for their own architectural judgment.

The framework is designed to address the dual challenge identified in the computing education literature: protecting academic integrity by ensuring AI cannot substitute for genuine learning (Ding, 2024; Akbar, 2025), while preparing students for an AI-augmented professional workforce (JetBrains, 2025; Modall, 2025).

References

Akbar, M. S. (2025). Beyond detection: Designing AI-resilient assessments with automated feedback tool to foster critical thinking and originality. University of Sydney. arXiv:2503.23622.

Ding, K. (2024). Designing AI-resilient assessments using interconnected problems: A theoretically grounded and empirically validated framework. University of Pennsylvania. arXiv:2512.10758.

JetBrains. (2025). State of developer ecosystem 2025. JetBrains Research.

Modall. (2025). AI in software development: 25+ trends and statistics. Retrieved from https://modall.ca/blog/ai-in-software-development-trends-statistics

Stack Overflow. (2025). Developer survey 2025. Stack Overflow.

GLOSSARY

Key Terms and Concepts

Terms are listed alphabetically. Chapter references in parentheses indicate where each term is introduced in depth.

Term	Definition
A	
abstract class	A class that names a concept shared by related classes without being instantiated itself. The word abstract is not used in this book's Java syntax, but the idea, that Student is the general concept from which GraduateStudent and ExchangeStudent inherit, is the foundation of Chapter 10's polymorphism work. *(Ch. 10)*
abstraction	One of the four computational thinking habits. Identifying what matters for a given problem and deliberately setting aside what does not. In object-oriented design, treating an object by its public interface rather than its internal implementation. *(Ch. 1)*
AI-Augmented	A workflow in which AI tools are used as a capable starting point for code generation, while the developer retains direction of intent and responsibility for design decisions. In this book, practiced through the optional Augment step of each Architectural AI Audit. *(Ch. 1)*
AI-Resilient	A habit of treating AI-generated output as a hypothesis to verify rather than a conclusion to accept. Practiced through the Audit and Verify steps of each Architectural AI Audit, and through the chapter-specific verification reminders in every practice problem section. *(Ch. 1)*
algorithm	A finite, ordered sequence of steps designed to accomplish a defined computational goal. Every program implements one or more algorithms. Decomposition is the skill of breaking a complex problem into smaller algorithms. *(Ch. 1)*
ArrayList	A built-in Java class that manages a resizable ordered list of objects. Grows and shrinks automatically. Key methods: add(), remove(), get(), set(), size(). Requires import java.util.ArrayList. Use ArrayList<Integer> for numbers, not ArrayList<int>. *(Ch. 14)*
array	A fixed-size, ordered collection of values of the same type, each reachable by its index. Positions start at 0; the last valid index is array.length - 1. Standard traversal: for (int i = 0; i < array.length; i++). *(Ch. 6)*
B	
base case	The stopping condition of a recursive method. When reached, the method returns a result directly without calling itself again. Every recursive method requires at least one base case, and every recursive call must make genuine progress toward it. *(Ch. 11)*

Term	Definition
binary search	A search algorithm that works on sorted arrays by repeatedly halving the search space. Compares the target against the middle element and eliminates half the remaining elements at each step. Runs in O(log n) time. Requires sorted input. *(Ch. 12)*
boolean	A data type with exactly two possible values: true or false. Controls conditional logic and loop conditions. Boolean expressions use relational operators (==, !=, <, >, <=, >=) and logical operators (&&, ||, !). *(Ch. 3)*
break	A statement that exits a loop or switch block immediately. Used in switch to prevent fall-through between cases. *(Ch. 7)*
C	
call stack	The internal record Java maintains of which methods are currently running and waiting for results. In recursion, grows with each recursive call and unwinds as each base case is reached. A StackOverflowError occurs when the stack grows too deep. *(Ch. 11)*
catch	A block that follows a try block and runs when a specific exception type is thrown. Names the exception type it handles in parentheses. Prevents the program from crashing and allows it to respond gracefully. *(Ch. 13)*
class	A blueprint describing what objects of a particular type know (fields) and can do (methods). A class definition does not create any objects; the new keyword creates objects from it. In this book: Student, GraduateStudent, and Roster are all classes. *(Ch. 7)*
compiler	The tool that translates Java source code into bytecode the Java Virtual Machine can execute. Also checks for syntax and type errors before the program runs. In the Professional Loop framework, the compiler is the Workshop: the neutral verifier of correctness, distinct from both AI and the student's judgment. *(Ch. 1)*
constructor	A special method that runs exactly once when a new object is created with new. Its name must match the class name exactly and it has no return type. Used to initialize an object's fields. *(Ch. 7)*
continue	A statement inside a loop that skips the rest of the current iteration and moves to the next one. Unlike break, it does not exit the loop. *(Ch. 7)*
CRC card	Class-Responsibility-Collaboration card. A design tool that captures three things about a class before any code is written: its name, the information it holds (fields), and the actions it can perform (methods). *(Ch. 7)*
D	
decomposition	One of the four computational thinking habits. Breaking a large, complex problem into smaller, individually solvable pieces. The first and most fundamental habit, applied in every chapter of this book. *(Ch. 1)*
do-while loop	A loop that runs its body at least once before checking its condition. Used when the loop must execute before the condition can be evaluated, most commonly for input validation: ask the user for input, then check whether it was valid. *(Ch. 13)*

Term	Definition
E	
encapsulation	A design principle that marks fields private and provides controlled access through public getter and setter methods. Allows validation logic to protect fields from invalid states, and lets a class change its internal implementation without breaking outside code. *(Ch. 8)*
exception	An object Java creates automatically when something goes wrong at runtime. If uncaught, it crashes the program and prints a stack trace. If caught with try/catch, the program can respond gracefully. Every exception type has a specific name identifying what went wrong. *(Ch. 13)*
extends	The keyword that declares an inheritance relationship. class GraduateStudent extends Student means GraduateStudent is a Student: it inherits all of Student's fields and methods. Always apply the is-a test before using extends. *(Ch. 9)*
F	
field	A variable declared inside a class but outside any method. Stores the state of an object. Each object has its own copy of its class's fields. Fields declared private can only be accessed through the class's own methods. *(Ch. 7)*
for loop	A loop that consolidates start, check, update, and body into one line. Preferred when the number of iterations is known in advance. Standard form: for (int i = 0; i < n; i++). This book uses i = i + 1 as a teaching scaffold; i++ is the professional standard. *(Ch. 4)*
for-each loop	An enhanced for loop that iterates over every element of an array or ArrayList without managing an index variable. Form: for (Type var : collection). Preferred when the index itself is not needed. *(Ch. 14)*
G	
getter	A public method that returns the value of a private field. Follows the naming convention getFieldName(). Provides controlled read access to encapsulated state. *(Ch. 8)*
I	
inheritance	A relationship where one class receives all fields and methods of another and can extend or override them. Declared with extends. Apply the is-a test to verify inheritance is appropriate before using it. *(Ch. 9)*
instanceof	An operator that tests whether an object is of a particular class, returning true or false. In well designed polymorphic code, instanceof checks are a design smell: a missing @Override usually makes the runtime type check unnecessary. *(Ch. 10)*
is-a relationship	The test for whether inheritance is appropriate. Ask: is every instance of the child class genuinely an instance of the parent class? GraduateStudent is always a Student (correct). Course is not a Student (architectural flaw, examined in Chapter 9's Architectural AI Audit). *(Ch. 9)*
L	

Term	Definition
linear search	A search algorithm that checks every element in order until the target is found or the array is exhausted. Returns the index of the target or -1 if not found. Works on both sorted and unsorted arrays. Runs in O(n) time. *(Ch. 12)*
M	
method	A named block of code inside a class that performs a specific action. Defines what an object can do. Declared with a return type (or void). Called with dot notation: objectName.methodName(). *(Ch. 7)*
method overriding	Replacing an inherited method with a new implementation in a child class. The overriding method must match the parent's name and parameters exactly. Always mark with @Override, which catches typos as compile-time errors rather than silent runtime bugs. *(Ch. 10)*
O	
object	A specific instance of a class, created with new. Holds its own copy of its class's fields and can call the methods defined in its class. The vending machine analogy: you use the public interface without needing to know the internal mechanism. *(Ch. 5)*
@	
@Override	An annotation placed above a method that is intentionally overriding an inherited one. Not required for overriding to work, but required in professional Java practice (Oracle and Google style guides). Turns a silent wrong-method bug into an immediate compiler error. *(Ch. 10)*
P	
parameter	A variable declared in a method signature that receives a value when the method is called. Distinct from an argument, which is the specific value passed at the call site. *(Ch. 7)*
pattern recognition	One of the four computational thinking habits. Identifying similarities, recurring structures, and shared patterns across problems. The skill that lets a solution to one problem become a template for a family of related problems. *(Ch. 1)*
polymorphism	The ability to call the same method on different object types and get behavior appropriate to each type automatically. Achieved through inheritance and @Override. No instanceof check needed: each object runs its own version of the method. *(Ch. 10)*
private	An access modifier that restricts a field or method to the class in which it is declared. The foundation of encapsulation. Code outside the class cannot read or write a private field directly. *(Ch. 8)*
public	An access modifier that makes a method or class accessible from any other class. Constructors and externally-called methods are declared public. Fields are almost always private. *(Ch. 7)*
R	
recursion	A technique where a method solves a problem by calling itself on a smaller version of the same problem. Requires a base case (where it stops) and a recursive case (where it calls itself with a smaller input). Analogous to Russian nesting dolls. *(Ch. 11)*

Term	Definition
recursive case	The part of a recursive method that calls itself on a smaller version of the problem. Must always make genuine progress toward the base case; if it does not, the method will produce a StackOverflowError. *(Ch. 11)*
return	A statement that immediately exits a method and optionally sends a value back to the caller. Methods with a non-void return type must return a value on every possible code path. *(Ch. 7)*
S	
selection sort	A sorting algorithm that repeatedly finds the smallest remaining element and swaps it into its correct position. Simple to trace by hand. Runs in O(n^2) time regardless of input order. *(Ch. 12)*
setter	A public method that updates the value of a private field, typically after validating the new value. Naming convention: setFieldName(newValue). Where business rules about valid field values belong. *(Ch. 8)*
StackOverflowError	A runtime error that occurs when a recursive method calls itself so many times that Java runs out of call stack space. Almost always caused by a missing or unreachable base case. *(Ch. 11)*
state	The collection of all variable values at a specific moment during a program's execution. Programs maintain state to track progress, accumulate results, or respond to user input. *(Ch. 2)*
static	A modifier that makes a method or field belong to the class itself rather than to any specific object. Static methods can be called without creating an object: Math.abs(-5). Most search and sort algorithms in this book are static. *(Ch. 12)*
String	A built-in Java class representing text. Has methods: length(), toUpperCase(), indexOf(), substring(), charAt(), equals(). Use equals() to compare String contents; == checks object identity, not text content. *(Ch. 2)*
super	A keyword that refers to the parent class. Used in a child class constructor to call the parent constructor: super(name, id, gpa) must be the first line. Also used to call a parent method that has been overridden. *(Ch. 9)*
switch	A selection structure that branches on the exact value of a single variable. Each branch is a case label. Always end each case with break to prevent fall-through into the next case. *(Ch. 7)*
T	
this	A keyword referring to the current object. Used in constructors to distinguish a parameter with the same name as a field: this.name = name assigns the parameter to the field. *(Ch. 9)*
try	A block that marks code that might throw an exception. If an exception occurs inside try, execution jumps to the matching catch block. Code after the exception-throwing line inside try is skipped. *(Ch. 13)*
V	

Term	Definition
variable	A named storage location in memory holding a value of a specific type. Declared with a type and a name: int count = 0. The type determines what operations are valid. *(Ch. 2)*
W	
while loop	A loop that checks its condition before each execution of the body. If false on the first check, the body never runs. Requires four ingredients: starting value, condition, update, and body. Standard when the number of iterations is not known in advance. *(Ch. 4)*
wrapper class	A class that wraps a primitive type as an object, enabling storage in an ArrayList and providing utility methods. Integer wraps int; Double wraps double. Key use: Integer.parseInt() converts a String to an int. Note: ArrayList<Integer>, not ArrayList<int>. *(Ch. 14)*

Index

Chapter references are abbreviated as Ch. followed by the chapter number. Page numbers will be finalized in the print edition.

A

B

C

D

E

O

P

R

S

T

V

W

www.ingramcontent.com/pod-product-compliance
Lightning Source LLC
LaVergne TN
LVHW070854210926
848482LV00016B/935